Beyond Compliance

The Refinery Manager's Guide to ISO 14001 Implementation

Beyond Compliance

The Refinery Manager's Guide to ISO 14001 Implementation

Nicholas P. Cheremisinoff, Ph.D.,

and Motasem B. Haddadin

Houston, Texas

Gulf Publishing Company
2 Greenway Plaza, Suite 1020
Houston, TX 77046

10 9 8 7 6 5 4 3 2 1

Printed in the United States of America.

Printed on acid-free paper.

Text design and composition by Ruth Maassen.

Library of Congress Cataloging-in-Publication Data

Cheremisinoff, Nicholas P.
Beyond compliance : the refinery manager's guide to ISO 14001 implementation / Nicholas P. Cheremisinoff and Motasem B. Haddadin.
p. cm.
Includes bibliographical references and index.
ISBN 0-9765113-9-8 (alk. paper)
1. ISO 14000 Series Standards. 2. Environmental protection—Standards. I. Haddadin, Motasem B. II. Title.
TS155.7.C4544 2006
658.4'083—dc22

2006010220

Contents

Preface

This volume is intended to assist refinery and environmental managers in implementing an environmental management system (EMS). For those facilities that already have an EMS, the concepts, practices, and tools provided in this volume should help strengthen their program.

A vast and rich library of resources are available in both printed and electronic form on the World Wide Web concerning EMSs; however, a great many of these resources discuss theory rather than practice. Additionally, tools that can aid in implementation are still evolving. The present volume does not attempt to synthesize the almost overwhelming literature nor does it focus on theory and concepts already presented by many others. Rather, the book provides a basic understanding of the necessity for an EMS, the steps applied to getting an EMS started and specific tools to use in an EMS.

The need for an EMS can be summed up in one simple phrase: It is simply not possible to stay in business very long without one. Environmental regulations and strict enforcement are widespread throughout the world, even in countries of transitioning and developing economies. Business globalization and the need to interact with multinational corporations, a complex and integrated supply chain, and a well-informed public have made EMSs crucial to both large and small corporations. In the United States, more so than any other country, an EMS is critical in the responsible management of ongoing environmental aspects that pose present and future financial threats to an organization and in defending against legal threats from past environmental damages. The courts are backlogged with third-party suits stemming from environmental damages caused by companies that focused on compliance only as opposed to improving environmental performance on a continual basis. While a cornerstone of any EMS is to comply with regulatory compliance obligations, the major intent is to implement a management tool that focuses a company's attention on continually improving environmental performance by eliminating all forms of waste and pollution at the source. By improving overall environmental performance in a systematic way and doing so continuously, a company's compliance record not only improves but operating costs are reduced. This achieves higher levels of profitability.

The focus of this monograph is the petroleum refining industry sector. The nature of petroleum refining and its downstream operations are such that it has a large number and variety of air, solid waste, and liquid effluent forms of pollution. Environmental, health, and safety regulations and local statutes governing pollution from refinery operations are complex and demanding. Even the best-run facilities face complex issues that are formidable challenges from the standpoint of meeting compliance obligations. And, with the increasing scarcity of crude oil, there is an acute need for refineries to become as efficient as possible.

The book is organized into five chapters and accompanied by two software programs contained on a single CD. Chapter 1 provides an overview of refining operations and the major sources of pollution. Discussions focus on the pollution and waste output streams. It is assumed that the reader already has a thorough background on the unit process operations used in oil refining; hence, process operations per se are not discussed.

Chapter 2 provides a condensed summary of the waste and pollution outflows from major unit process operations used in a typical refinery. The information is organized to assist the reader in identifying environmental aspects that require management through an EMS.

Chapter 3 provides a concise explanation of what an environmental management system is. Topics covered are the general concepts of an EMS, EMS standards, BS 7750 (the British EMS standard), EMAS (the European Union's Eco-Management and Audit Scheme), ISO 14000 and in particular the ISO 14001 EMS, comparisons among the various standards, the difference between compliance and conformance with environmental performance standards, and finally the principle drivers for adopting an EMS.

Chapter 4 explains the components of an EMS. Topics covered include creating environmental policy, the initial environmental review (IER), environmental aspects (EA), meeting legal requirements, developing an environmental management program (EMP), how to implement environmental policy, measuring and evaluating performance, conducting EMS audits and management reviews, and reporting and communication tools.

Chapter 5 discusses tools for conducting the initial environmental review, or IER. An IER is among the first crucial steps to implementing an EMS. The chapter first provides a detailed overview of major U.S. environmental regulations. The balance of the chapter covers the application of the two software products contained on the CD.

There are two programs on the accompanying CD: IER (Initial Environmental Review) Auditor and Environmental Aspects (EA) Evaluator. No special requirements are needed to use the programs other than Microsoft Access installed on a PC. Both programs must be downloaded onto the computer before becoming operational.

IER Auditor is a multipage audit questionnaire prepared as a database management tool using Microsoft Access as the platform. There are 11 files or spreadsheets to gather information on. The program facilitates data collection and information collation from an IER audit. Application of this software tool enables the facility to rapidly identify areas of noncompliance as well as parts of the business where significant negative environment impacts are likely to exist. The user can begin developing environmental strategies for corrective action based on a review of the information acquired from the audit. The software enables the user to generate a report that provides a baseline characterization of how the refinery manages its environmental aspects and how well or poorly it meets their regulatory and other requirements. IER Auditor contains a Tutorial to guide the user through the operational features of the program. It is recommended that the reader first read Chapter 5.

Once the information has been documented and general strategies developed, the environmental manager, along with members of the IER audit team, can begin acquiring more-detailed information on specific environmental aspects and their impacts. Using EA Evaluator, it is possible to rank the priorities of the significant environmental aspects. After this has been accomplished, effort can get underway to develop environmental action plans aimed at corrective actions. The intent of the software is to assist the environmental manager in ranking the priorities of environmental aspects. This enables recommendations to be devised on strategy and scheduling of corrective actions. EA Evaluator contains a Tutorial that explains and illustrates the operational features of the program. When opening EA Evaluator, go directly to the Tutorial page and read through the step-by-step instructions. It is recommended that the reader first read Chapter 5.

For over 20 years, I have worked with the Gulf Publishing team. As always, they diligently worked to produce high-quality reference materials that service the industry and academic communities. I and my coauthor wish to extend our deepest gratitude for their efforts in the production of this volume and accompanying software tools.

Nicholas P. Cheremisinoff
Harpers Ferry, WV

Motasem B. Haddadin
Amman, Jordan

About the Authors

Nicholas P. Cheremisinoff has 30 years of international consulting and project management experience dealing with environmental management practices and technologies. His assignments focused on corporate responsible care through the application of pollution prevention. He served both as a practitioner and educator to industry stakeholders on responsible care practices that strive to improve environmental performance through the application of cost-effective clean technologies. He designed and implemented projects for the United States Agency for International Development, the World Bank Organization, the European Union, the U.S. Trade and Development Agency, the U.S. Department of Energy, the U.S. Department of Defense, and a number of Fortune 100 and multinational companies. He applied his experiences and knowledge in environmental toxicity tort litigations as an expert witness in the fields of environmental forensics, historical reconstruction of environmental management practices, and formulating opinions regarding levels of responsible care. Dr. Cheremisinoff is the author, coauthor, or editor of more than 150 books and numerous state-of-the-art review articles and R&D papers. He received his B.S., M.S., and Ph.D. degrees in chemical engineering from Clarkson College of Technology.

Motasem B. Haddadin holds a Masters degree in Water Resource and Environmental Engineering from Jordan University of Science and Technology. His specialty areas include water resource and water demand management, drought assessment and management using GIS systems, and software engineering, which includes system design and programming. Mr. Haddadin has worked on international water conservation and pollution prevention programs funded by the U.S. Agency for International Development as well as for private sector companies and the Jordanian Ministry of Environment and the national Water Authority.

1

Refining Operations and the Sources of Pollution

Introduction

In today's world, *responsible* companies adhere in one form or another to an environmental management system standard. While small enterprises may not need to formalize a program, they are no less vulnerable than large multinational corporations to liabilities stemming from mismanagement of environmental obligations and hence need to rely on the same structured management system and approach to meeting their compliance obligations. Because of the globalization of markets and linkages through worldwide supply chains, it is not possible to conduct business without having market demands impose the need to adopt elements of, if not full, environmental management systems to compete as a preferred supplier. In short, an environmental management system is a 21st century necessity or, perhaps better said, a requirement, for doing business.

Petroleum refining is among the leading manufacturing industries in the United States, especially in terms of its share of the total value of shipments of the American economy. And, while there are a few number of facilities and companies worldwide, the impact on global economy is perhaps unmatched by any other sector.

Petroleum refining is the physical, thermal, and chemical separation of crude oil into major distillation fractions. These factions are further processed through separation and conversion steps into finished petroleum products. Products fall within three major categories: fuels, which include motor gasoline, diesel and distillate fuel oil, liquefied petroleum gas, jet fuel, residual fuel oil, kerosene, and coke; finished nonfuel products, which include solvents, lubricating oils, greases, petroleum wax, petroleum jelly, asphalt, and coke; and chemical industry feedstock, which includes naphtha, ethane, propane, butane, ethylene, propylene, butylenes, butadiene, benzene, toluene, and xylene. About 40% or more of the total energy consumed in the United States is derived from petroleum products on the basis of BTUs consumed. The chemical industry feedstocks are primary inputs to a vast number of products that range from fertilizers, pesticides, waxes, thinners, solvents, cleaning fluids, detergents, refrigerants, antifreeze, paints, sealants, cleaning solutions, insulations, latex, rubber compounds, hard plastics, plastic sheeting materials, plastic foam, and synthetic fibers, to name a few. Nearly 90% of the petroleum products used in the United States are fuels, with motor gasoline accounting for roughly 43% of the total.[1]

The nature of petroleum refining and its downstream operations is such that there are a large number and variety of air, solid waste, and liquid effluent forms of pollution. Environmental and health and safety regulations and local statutes governing pollution from refinery operations are complex and demanding. Even the best-run facilities face complex issues that are formidable challenges from the standpoint of meeting compliance obligations. And, because many facilities throughout the United States, if not the world, are mature operations, they have the legacy of past environmental damages from eras when the level of responsible care toward the environment and public safety were limited by technologies and understanding of fate and transport mechanisms for pollution of the day.

[1]*Profile of the Petroleum Refining Industry*, EPA Office of Compliance Sector Notebook Project, EPA 310-R-95-013 (Washington, DC: U.S. Environmental Protection Agency, 1995).

This first chapter provides orientation on refinery operations and the sources of pollution and waste. Responsible management of these byproducts both from a regulatory standpoint and in terms of optimum economic performance of the business is the primary objective of an environmental management system. Subsequent chapters present the reader with the theory, applications, and tools for implementing an environmental management system.

1.1 The Nature of Pollution

It is assumed that the reader already has a general, if not thorough, understanding of the major industrial processes within the petroleum refining sector, including the materials and equipment used and the major processes employed. This section and the remainder of the chapter provide a general understanding of the interrelationship between the industrial processes, pollutant outputs, and pollution prevention opportunities. A list of recommended references is provided at the end of the chapter for those readers desiring more in-depth information.

Crude oil is best described as a mixture of hydrocarbons with small quantities of impurities. Its composition varies significantly depending on the source. The consequence of this is that refineries in general are complex systems comprising multiple operations. The specific operations and technologies relied on for refining are a function of the composition and properties of the crude oil to be refined and the desired products. It is fair to state that no two refineries throughout the world are exactly alike.

The process of refining can be separated into two phases and a number of supporting operations and technologies. The first phase, known as *desalting*, is followed by distillation of the crude into various components or "fractions." The second phase comprises three downstream processes, known as *combining*, *breaking*, and *reshaping*.

Downstream processes are responsible for the conversion of some of the distillation fractions into useful products, such as residual fuel oil, gasoline, and kerosene. This is accomplished through combinations of cracking, coking, reforming, and alkylation processes.

Supporting process operations may include wastewater treatment, sulfur recovery, additive production, heat exchanger cleaning, blowdown systems, product blending, and product storage. A brief description of each general process follows, with focus given to the pollution outputs.

1.2 Desalting

This operation is applied before the crude oil is separated into fractions. Desalting is a treatment process that removes corrosive salts along with some of the metals and suspended solids, which can cause catalyst deactivation.

The process involves mixing water with heated crude oil. As a rule of thumb, the amount of water added to the mixture is between 3% and 10% of the crude oil volume. This range provides a sufficient concentration of water to dissolve the salts into the water.

Once the salts are dissolved, the water fraction must be separated from the crude. In older refineries, this is accomplished in a separating vessel with the addition of demulsifier chemicals that aid in breaking the emulsion. A more-common technology used is electrostatic coalescence. This involves applying a high potential electric field across the settling vessel to coalesce the polar saltwater droplets.

The desalting process generates an oily sludge along with a high temperature saltwater waste stream. The saltwater waste stream is typically added to other process wastewaters for treatment.

From a pollution-prevention standpoint, the water used in crude desalting is often untreated or only partially treated and is derived from other refining-process water sources.

1.3 Distillation

Distillation involves the heating, vaporization, fractionation, condensation, and cooling of feedstocks. There are two basic distillation processes: atmospheric and vacuum distillation.

1.3.1 Atmospheric Distillation

In atmospheric distillation, the desalted crude is heated in a heat exchanger and furnace to approximately 750°F then fed to a vertical distillation column under pressure. The crude vaporizes and separates into various fractions by condensing on 30–50 fractionation trays, each corresponding to a different condensation temperature. Lighter fractions condense and collect toward the top of the column. Heavier fractions, which may not vaporize in the column, undergo further separation in a second column by vacuum distillation.

Within each atmospheric distillation column are four or more sidestreams of low-boiling components that are removed from the tower from different trays. The low-boiling-point mixtures are in equilibrium with the heavier fractions that must be removed. Each of the sidestreams is sent to a different small stripping column. The small stripping tower contains 4–10 trays, with steam injected under the bottom tray. Steam is used to strip light-end components from the heavier components, and both the steam and stripped components are fed back to the atmospheric distillation column above the corresponding sidestream draw tray. The fractions obtained from the atmospheric column include naphtha, gasoline, kerosene, light fuel oil, diesel oils, gas oil, lube distillate, and heavy bottoms. The majority of these can be sold as finished products or blended with products derived from downstream processes.

Another product that can be produced from an atmospheric distillation tower is light, noncondencible refinery fuel gas, which is mainly methane and ethane. This gas contains hydrogen sulfide and ammonia. The mixture of these gases is referred to as *sour gas* or *acid gas*. Sour gas is sent to a sour gas treatment system, which separates the fuel gas so that it can be used as fuel in the refinery heating furnaces. The sulfur can be recovered in this process as elemental sulfur or for the production of sulfuric acid.

Significant air emissions result from atmospheric distillation due to the combustion of fuels in the furnaces used to heat the crude, from process vents, and from a variety of fugitive emission sources. Also, oily sour water, which is derived from condensed steam containing hydrogen sulfate, ammonia, and oil, is produced from the fractionators.

1.3.2 Vacuum Distillation

Vacuum distillation is used to fractionate the heavier fractions from the atmospheric distillation column that cannot be distilled without cracking under its pressure and temperature conditions. Figure 1–1 shows a photo of a unit in operation in the Middle East.

Vacuum distillation is the distillation of petroleum factions at low pressure (typically 0.2–0.7 psia), which increases volatilization and subsequent separation. The vacuum inside the fractionator is maintained using steam ejectors and vacuum pumps, barometric condensers or surface condensers, or a combination thereof. Injection of superheated steam at the base of the vacuum column further

Figure 1–1 A vacuum distillation unit operating in the Middle East.

reduces the partial pressure of the hydrocarbons, thereby facilitating vaporization and separation. Heavier fractions within the column are processed downstream, thus creating higher-valued products either by the methods of cracking or coking.

Sources of emissions are the combustion of fuels in the furnace along with some light-end gases, leaving the top of the condensers on the vacuum distillation column. Some of the noncondensable light hydrocarbons and hydrogen sulfide pass through the condenser to a hot well, from which they are discharged to the refinery sour fuel system or vented to a process heater, a flare, or some other control device to destroy hydrogen sulfide. The quantity of emissions is influenced by the unit's size, the type and nature of the feedstock, and the cooling water temperature. When barometric condensers are used in combination with a vacuum distillation column, significant amounts of oily wastewater can be generated. Modern refinery operations have implemented pollution prevention programs whereby vacuum pumps and surface condensers replaced barometric condensers to eliminate or reduce the oily wastewater stream. Oily sour water is also a byproduct in the fractionators.

1.4 Downstream Processing Operations

A large number of downstream processing technologies are used to create higher-valued products. The principle technologies are

- Thermal cracking (known as *visbreaking*)
- Coking
- Catalytic cracking
- Catalytic hydrocracking
- Hydrotreating
- Alkylation
- Isomerization
- Polymerization
- Catalytic reforming

- Solvent extraction
- Merox
- Dewaxing
- Propane deasphalting

These processes alter the molecular structure of hydrocarbon molecules either by the breaking of bonds to form smaller molecules, joining structures to create larger molecules, or reshaping structures to form higher-value molecules. A large number of variations of techniques are used throughout the industry, some of which are proprietary in nature.

1.4.1 Thermal Cracking or Visbreaking

Thermal cracking, also known as *visbreaking*, is an older process that capitalizes on heat and pressure to break large hydrocarbon molecules into smaller, light molecules. The more-modern and more-efficient technology is catalytic cracking. Still, thermal cracking is an important technology, and some smaller, older units in parts of the world rely on it. Both processes aim to convert less-valuable products, such as heavy fuel oil and cutter stock, into feedstock that feeds the catalytic reactor for higher-valued products and higher gasoline yields.

The feedstock is heated in a furnace to a temperature of up to 1000°F then fed to a reaction chamber maintained at a pressure of around 140 psig. After the reactor stage, the process stream is mixed with a cooler recycle stream to quench the cracking reactions. The term *quenching* refers to the cooling of the reaction, thus bringing about termination. The product then flows to a flash chamber where the pressure is let down (i.e., reduced), causing the lighter products to vaporize. These lighter fractions may be recovered and sent to a fractionating tower, where the various fractions are separated.

The "bottoms" consist of heavy residue. Part of this residue is recycled to cool the process stream leaving the reactor chamber. The remaining bottoms can be blended into residual fuels.

Sources of thermal-cracking air emissions are the combustion of fuels in the process heater, vents, and fugitive emissions. A sour water stream is generated in the fractionator.

1.4.2 The Coking Process

The process of coking is used to reduce refinery production of low-value residual fuel oils. Its role is to convert these low-value products to transportation fuels (e.g., gasoline and diesel). Coking also produces petroleum coke (or petcoke).

Petcoke is solid carbon with varying amounts of impurities. It has long been a source of fuel for power plants when the coke product has a low sulfur content. Most recent commercial projects use refinery waste or products that have marginal market value, a category into which petcoke and some heavy oils fall. These projects are referred to as *trigeneration* plants, because they produce hydrogen, power, and steam for use within the refinery and for export. These projects rely on gasification technologies. Recent gasification projects in the United States using petcoke or heavy oils as feedstock include the Frontier Oil gasification project in El Dorado, Texas; Motiva gasification project in Delaware City, Delaware; Farmland Industries gasification project in Coffeyville, Kansas; and the Exxon-Mobile gasification project in Baytown, Texas. European refinery gasification projects include the API project in Falconara, Italy; Sarlux project in Sardinia, Italy; ISAB project in Sicily, Italy; and Shell project in Pernis, the Netherlands.

Coke also has nonfuel applications. It has been used traditionally as a raw material for many carbon and graphite products, including anodes for the production of aluminum and furnace electrodes for the production of elemental phosphorous, titanium dioxide, calcium carbide, and silicon carbide.

There are several technology routes to making coke. The two most common are "delayed coking" and "fluid coking." Delayed coking is traditional and still the most widely used. Fluid coking produces higher grades of coke.

Delayed coking relies on the same basic process as thermal cracking, with the exception that feed streams are allowed to react longer without undergoing quenching. The delayed coking feed streams are derived from residual oils from different upstream processes. These are sent to a fractionating tower, where the residual lighter fractions are drawn off and the heavier ends condensed. The heavier ends are removed and heated in a furnace then fed to an insulated vessel, known as a *coke drum*, where coke is formed. When the coke drum becomes filled with product the feed is diverted to another coke drum that is in parallel. Hot vapors from the coke drums contain cracked lighter materials, hydrogen sulfide, and ammonia. These are recycled back to the fractionating column, where they can be treated in the sour-gas treatment system or recovered as intermediate products as sidestreams from the fractionator. Steam is injected into the full coke drum to remove or strip hydrocarbon vapors. Water is injected to quench the coking reactions, allowing the product to be finished and removed.

The air emissions include the process heater flue gas, fugitive emissions, and emissions from the removal of coke from the coke drum. Injected steam is condensed, and the remaining gases are sent to a flare. Wastewater is also generated during the coke removal, steam injection, and quenching steps. The coke recovery step from the drum also releases particulate emissions and residual hydrocarbons as vapors.

1.4.3 The Catalytic Cracking Process

The catalytic cracking process uses heat, pressure, and a catalyst to break larger hydrocarbon molecules into smaller, lighter molecules of greater value. Catalytic cracking has long replaced thermal cracking in modern refineries, because it produces more gasoline with a higher octane and less heavy fuel oils and light gases.

The feedstocks comprise light and heavy oils derived from the crude oil distillation tower. They are processed primarily into gasoline, fuel oil, and light gases. The catalysts used in the process are mixtures of crystalline synthetic silica-alumina (called *zeolites*) and amorphous synthetic silica-alumina. The catalytic cracking process produces coke as a byproduct, which collects onto the surfaces of the catalyst. This form of fouling reduces the effectiveness of the catalyst. The catalyst therefore needs to be regenerated continuously or on a frequent and periodic basis, which is accomplished by burning the coke off of the catalyst surfaces at high temperatures. The particular method and frequency of catalyst regeneration are major considerations in the design of catalytic cracking units. The competing technologies include the use of fixed-bed reactors, moving-bed reactors, fluidized-bed reactors, and single-pass or once-through reactor systems. Fluidized-bed and moving-bed systems are among the most widely used.

Fluidized-bed cracking units are commonly referred to as FCCUs. In this process, oil and oil vapor are preheated (typically to between 500°F and 800°F) then contacted with hot catalyst at a temperature of about 1300°F. The contacting of the streams may be accomplished in either the reactor itself or in the riser section (i.e., the feed line) to the reactor. The catalyst is a relatively fine, granular material that, when mixed with the hot vapors, acquires many of the same macroscopic properties of a fluid. The fluidized catalyst and the reacting hydrocarbon vapors readily separate in the reactor, and any oil remaining on the catalyst is removed by applying steam stripping. Cracked oil vapors are fed to a fractionating column, where different desired fractions are separated and recovered. The catalyst is diverted into a separate vessel for single- or two-stage regeneration, which is accomplished by burning off the coke deposits with hot air.

In contrast, the moving-bed process first heats the oil to about 1300°F. The hot oil is passed under pressure through the reactor, where it is contacted with a catalyst in the form of tiny beads or pellets. Cracked products then are passed to a fractionating tower, where different streams are separated and recovered. The catalyst is regenerated in a continuous process, where the deposits of coke are burned off. Steam can be used to strip the remaining hydrocarbons and oxygen from the catalyst surface before being fed back to the incoming oil stream.

Catalytic cracking units are a major source of air emissions. The air emission sources include the process heater flue gas, fugitive emissions, and regeneration of the catalyst. High concentrations of carbon monoxide (CO) are generated during the catalyst regeneration step. CO is converted to carbon dioxide (CO_2) in the regenerator and also further downstream in the carbon monoxide waste heat boiler. Significant amounts of fine catalyst dust are generated in FCCUs as a result of attrition from the constant movement and collisions between catalyst particles. Much of the dust is composed of alumina and small amounts of nickel. These are carried out with the CO stream to the CO burner. Catalyst dust is separated from the CO_2 stream using mechanical separators like cyclones or electrostatic precipitators. The waste generated here is a regulated hazardous waste that requires off-site transport and disposal.

Generated wastewater is composed of sour water from the fractionating column. This stream contains oils and phenols. Wastewater containing metal impurities from the feed oil is also produced from the steam injection step used to purge and regenerate the catalyst.

1.4.4 The Catalytic Hydrocracking Process

Catalytic hydrocracking uses a fixed-bed cracking reactor that operates at 1200 to 2000 psig in the presence of hydrogen. Feedstocks sent to hydrocracking units are those fractions that are difficult to crack and cannot be cracked efficiently in catalytic cracking units. The streams include middle distillates, residual fuel, cycle oils, and reduced oils.

Hydrogen suppresses the formation of heavy residual material and increases the yield of gasoline by reacting with the cracked products. The process also breaks the heavy sulfur- and nitrogen-bearing hydrocarbons, releasing these impurities to where they could foul the catalyst. Therefore, the feedstock first must be hydrotreated to remove impurities before being fed to the catalytic hydrocracker. Hydrotreating can also be accomplished by using the first reactor of the hydrocracking process to remove the impurities.

Water is a catalyst poison and must be removed from the feed stream. Water is removed by passing the feed stream through silica gel or a molecular sieve dryer. Catalytic hydrocracking can be either a single- or two-stage process, depending on the desired products and the production capacity. The catalysts used are usually a crystalline mixture of silica-alumina with small amounts of rare earth metals.

Feedstocks to the hydrocracking unit are initially hydrotreated to remove hydrogen sulfide and ammonia, as these are catalyst poisons. Sour gas and sour water streams are byproducts generated at the fractionator; however, when the hydrocracking feedstocks are first hydrotreated to remove the impurities, both streams will contain low levels of hydrogen sulfide and ammonia. Catalysts are usually regenerated off-site after about two to four years of operation. The consequence of this is that few emissions are generated from the regeneration process. The primary air emission sources are the process heater, vents, and fugitive.

1.4.5 The Hydrotreating and Hydroprocessing Processes

Hydrotreating and hydroprocessing are similar processes. They are designed to remove specific impurities such as sulfur, nitrogen, oxygen, halides, and metals, which act as catalyst poisons.

Hydrotreating is also used to upgrade the quality of fractions by converting olefins and diolefins to paraffins to reduce gum formation in fuels. Hydroprocessing uses residuals from the crude distillation columns, cracking these heavier molecules to lighter, more-valuable products.

Hydrotreating and hydroprocessing units are placed upstream of those processes in which sulfur and nitrogen could adversely affect catalysts (e.g., catalytic reforming and hydrocracking units).

Both processes use catalyst in the presence of large amounts of hydrogen under high pressure and temperature to react the feedstocks and impurities with hydrogen. Reactors are of the fixed-bed type, which require turnarounds for catalyst replacement on the order of months to years.

In addition to the treated products, the processes generate a stream of light fuel gases, hydrogen sulfide, and ammonia. Treated product and hydrogen-rich gas are cooled on exiting the reactor then separated. The hydrogen is recycled back to the reactor.

Off-gas streams are generally rich in hydrogen sulfide and light fuel gas. The fuel gas and hydrogen sulfide are sent to the sour gas treatment and sulfur recovery units. The catalyst systems used are typically cobalt and molybdenum oxides of alumina; however, some systems contain nickel and tungsten.

The air emissions sources from hydrotreating are the process heater flue gas, vents, and fugitive.

1.4.6 Alkylation

Alkylation is the process used to produce high-octane gasoline, blending stock from isobutene, which is formed largely during catalytic cracking and coking operations, and to a lesser extent from catalytic reforming, crude distillation, and natural gas processing. The process involves joining an olefin and an isoparaffin compound, using either sulfuric acid or a hydrofluoric acid catalyst system. Products include alkylates, including butane and propane liquids. At an acid concentration of about 88%, some of the acid must be removed and replaced with a stronger acid. In the hydrofluoric acid process, the slipstream of acid is redistilled. Any dissolved polymerization byproducts are removed from the acid stream. These are generally thick, dark-colored oils. Concentrated hydrofluoric acid is recycled. Hydrofluoric acid alkylation units are sophisticated and demand special engineering and operator training/experience. They require special safety equipment precautions to protect operators from exposure to the highly corrosive hydrofluoric acid.

In the sulfuric acid process, the acid removed must be regenerated in a sulfuric acid plant. The sulfuric acid plant is usually not part of the alkylation unit and most often is located off-site. Hence, special handling and on-site storage considerations are required. The air emission sources include process vents and fugitive.

1.4.7 Isomerization

Isomerization is a process that alters the molecular arrangement of the products without adding or removing anything from the original molecule. Paraffins, such as butane and pentane, from the crude distillation column are converted to isoparaffins having higher octane. Reactions take place in the presence of a platinum catalyst impregnated on a base material and at temperatures in the league of 200–400°F.

The isomerization catalyst system requires continuous addition of small amounts of organic chlorides, which convert to hydrogen chloride in the reactor. The feed stream needs to be free of oxygen and moisture (water is a source of oxygen) to avoid catalyst deactivation and corrosion problems. A competing catalyst system relies on a molecular sieve base and does not require a dry, oxygen-free feed stream. Both catalyst systems require a hydrogen environment to minimize coke deposits, which can cause fouling. Catalysts require replacement every two to three or more years. When replaced, the platinum is recovered from the spent catalyst. This is accomplished off-site by the supplier.

Light ends are stripped from the product stream exiting the reactor and sent to the sour gas treatment unit. There are several variations of treatment processes. A common one is to utilize caustic treating of the light fuel gas stream. This neutralizes any hydrochloric acid entrained in the stream. However, this approach does produce calcium chloride and other salts in the waste stream. Sources of air emissions include the process heater, vents, and fugitive. Wastewater streams contain caustic wash and sour water.

1.4.8 Polymerization

Polymerization is used to convert propene and butane to high-octane gasoline blending stock. The process resembles alkylation in its feed and products. Generally, it is considered a less-expensive alternative to alkylation.

Reactions take place under high pressure in the presence of a phosphoric acid catalyst system. This requires a feed that is free of sulfur and oxygen. Sulfur serves as a catalyst poison by neutralizing it. Oxygen slows the reactions and creates unwanted byproducts.

Propene and butane feed is washed with caustic to remove mercaptans. Mercaptans are relatively small molecules containing sulfur. The washing step is followed by a second washing with an amine solution to remove hydrogen sulfide. A third washing step involves water to remove caustics and amines. The feed then is dried by passing it through silica gel or a molecular sieve dryer.

Air emission sources that include sulfur dioxide come from the caustic washing stage. Spent catalyst is disposed of as a solid waste as it cannot be rejuvenated. Wastewater streams contain sour water with amines and mercaptans along with the caustic wash.

1.4.9 Catalytic Reforming

The catalytic reforming process uses catalytic reactions to convert low-octane, heavy, straight-run gasolines and naphthas from the crude distillation unit into high-octane aromatics, among which is benzene. The major reactions in this process are

- Dehydrogenation of naphthenes to aromatics
- Dehydrocyclization of paraffins to aromatics
- Isomerization
- Hydrocracking

Dehydrogenation reactions are endothermic, and as such, the hydrocarbon stream must be heated between each catalyst bed. The hydrocracking reaction is the only one that does not release hydrogen. Hydrogen can be used in the hydrotreating and hydrocracking processes.

Systems comprise fixed-bed or moving-bed configurations arranged in a series of three to six reactors. Feedstocks sent to the catalytic reforming process are first hydrotreated to remove sulfur, nitrogen, and metals.

In a continuous reforming process, the catalyst can be regenerated one reactor at a time and usually one to two times per day with no downtime to production. In semiregenerative systems, regeneration of all the reactors is carried out simultaneously after several months (up to 24 months with some processes); however, a turnaround is required. Reformulated gasoline rules have limited benzene in gasoline, so catalytic reforming has been systematically phased out as an octane enhancer process.

Air emission sources include process heater gas and fugitive emissions, with relatively small emissions coming from catalyst regeneration (regeneration is performed off-site). Small volumes of wastewater containing sulfides, ammonia, and mercaptans are derived from the stripping tower used to remove light ends from the reactor effluent stream.

1.4.10 Solvent Extraction

Solvent extraction employs solvents to dissolve and remove aromatics from lube oil feedstocks to improve viscosity, enhance oxidation resistance, improve color, and reduce gum formation. A wide variety of solvents can be used, but the two most common ones are furfural and phenol.

In a solvent extraction process, the feed lube stocks are brought into contact with the solvent in a packed column or rotating disc contactor. Different solvents have different solvent-to-oil and recycle ratios within the column. Solvent recovery from the oil stream is accomplished through distillation and steam stripping in a fractionator. The stream extracted from the solvent generally contains high concentrations of hydrogen sulfide, aromatics, naphthenes, and other hydrocarbon constituents. These are usually fed to the hydrocracking unit. The water stream from the fractionator contains oils and solvents.

1.4.11 Chemical Treating

Chemical treating is used to remove or change undesirable properties resulting from sulfur, nitrogen, or oxygen-compound contaminants in final petroleum products. The two basic processes are extraction and oxidation. Oxidation is referred to as *sweetening*. The selection of the process depends on the product.

Extraction removes sulfur from very light petroleum fractions. Examples are propane and propylene (PP) and butane and butylenes (BB). Sweetening is more effective on gasoline and middle-distillate products.

The "merox" extraction process removes mercaptans from PP and BB streams. PP streams undergo amine treating prior to the merox process to remove excess hydrogen sulfide. Hydrogen sulfide tends to fractionate with PP and interferes with the extraction process. A caustic prewash of the PP and BB streams removes remaining traces of hydrogen sulfide prior to the extraction step. Figures 1–2 and 1–3 show a caustic preparation operation.

PP and BB streams are passed up through the trays of an extraction column. Caustic solution flowing downward through the

Figure 1–2 Caustic preparation stage. This example shows material mishandling that can lead to site contamination and material losses, which cost money.

Figure 1–3 Shows mixing and metering stage of caustic stream preparation.

column absorbs mercaptans for the streams. The rich caustic is regenerated by oxidizing the mercaptans to disulfide in the presence of aqueous merox catalyst, while the lean caustic is recirculated to the extraction column. Since the disulfide is insoluble in the caustic, it can be separated out.

Sweetening is used on gasoline and distillate fractions. One common process is also a merox process that relies on a solid catalyst bed. Air and a small amount of alkaline caustic are injected into the hydrocarbon stream. As the hydrocarbon passes through the catalyst bed, the mercaptans are oxidized to disulfide. The caustic is not regenerated in this process. The disulfide remains with the gasoline, since it does not possess the obnoxious odor properties of mercaptans; hence, the term *sweetening* applies.

In extraction, a waste oil disulfide stream is generated in the separator. Air emission sources are fugitive in nature and process vents on the separator that contain sulfides.

1.4.12 Dewaxing

Lubricating oil base stocks are dewaxed to ensure that the product has the desired viscosity at lower ambient temperatures. There are two processes: selective hydrocracking and solvent dewaxing.

Selective hydrocracking uses zeolite catalysts to selectively crack wax paraffins. Solvent dewaxing dilutes the oil feed with a solvent to lower the viscosity. The mixture then is chilled to the point where the wax crystallizes and either settles out of solution or can be filtered. Solvents used for this purpose are methyl ethyl ketone (MEK) with methyl isobutyl ketone (MIBK) and MEK with toluene. The solvent is recovered from the oil and wax by the sequentially executed processes of heating, two-stage flashing, and steam stripping. The solvent recovery stage produces solvent contaminated in water as a waste. This stream is sent to the wastewater treatment plant. The wax can be reused as feed to the catalytic cracker or deoiled and sold into low-valued markets as an industrial wax. Air emission sources are fugitive and contain solvents.

1.4.13 Propane Deasphalting

The process of propane deasphalting produces lubricating oil base stocks by extracting asphaltenes and resins from residuals of the vacuum distillation column. Propane is used to remove asphaltenes due to its unique solvent properties. At low temperatures (100–140°F), paraffins are soluble in propane; and at higher temperatures (200°F), all hydrocarbons are nearly insoluble in propane. The process resembles solvent extraction with equipment that is comparable if not identical. As with solvent extraction, propane deasphalting uses a packed or baffled extraction column or a rotating disc contactor to mix the oil feed stocks with the solvent. When a countercurrent operating column is employed, four to eight volumes of propane are fed to the bottom of the column for every volume of feed fed to the top of the column. The oil is more soluble in the propane and therefore dissolves and flows to the top of the column. The asphaltenes and resin flow to the bottom where they are removed in a propane mix. Propane is recovered from the two streams using a flash system followed by steam stripping, in which propane is condensed and removed by cooling at high pressure in a first stage followed by low pressure in a second stage.

Figure 1–4 Small asphalt blending plant operation.

The recovered asphalt can be blended with other asphalts or heavy fuels or it can be used as feed to the coker. Figure 1–4 shows a small asphalt blending operation.

The propane recovery stage produces propane contaminated water, which is sent to the wastewater treatment plant. Air emissions are from fugitive sources and process vents.

1.5 Supporting Operations and Technologies

Refineries are complex manufacturing facilities that can be designed to produce a wide variety of products. The main production processes are aided by a large number of supporting operations, which range in function from waste and pollution management to product storage to heat exchange to final product modification and quality control. The following provides an overview of the major supporting operations.

1.5.1 Wastewater Treatment Processes

Large amounts of wastewater are generated by refinery operations. There are four major sources: surface water runoff, cooling water, process water, and sanitary wastewater.

Surface water runoff is intermittent and contains the constituents from spills onto ground and pavement areas, leaks from equipment, materials that collected in drains, and water brought in from the crude and product storage tanks roof drains along with precipitation wash off from operating units.

A significant amount of the water in a refinery is used for cooling purposes. Cooling water generally does not come into direct contact with process oil and product streams, but poor maintenance practices can lead to contamination. Even with well-maintained operations, some contamination will exist, but there will be less contaminants than process wastewater. For efficiency and cost savings reasons, most cooling water is recycled many times through the plant with a bleed or blowdown stream to the wastewater treatment works. Blowdown is necessary to control the buildup in concentration of contaminant and solids in the water. Solids buildup results in scale formation and can permanently damage heat exchange surfaces, thereby reducing thermal efficiency and adding to pollution in the noncontact water streams because of corrosion problems. Cooling towers, discussed later, exist within the recycle loop and are designed to cool the water using ambient air or with refrigeration systems. Very few plants today operate on a "once through" basis, where cooling water is passed through a process unit once then discharged directly without treatment to the wastewater treatment works.

Cooling water generally contains chemical additives. These additives are typically phosphates and antifouling biocides used to prevent exchanger surface and pipe scaling and to control biological growth, which is also a fouling agent. Older refineries used toxic chromates; however, it is unlikely these are used to any extent today. While the intent is not to allow cooling water to come into contact with oil process streams, poor maintenance and housekeeping caused by leaks and spills inevitably causes contamination.

The water used in process operations represents a large portion of any refinery's water balance. Process wastewater streams emanate from desalting operations, steam stripping, pump gland cooling, product fractionator reflux drum drains, and boiler blowdown operations. Since process waters come into direct contact with oil, the waste streams are highly contaminated.

Refinery wastewater treatment plants are integrated systems that use both primary and secondary treatment. Primary treatment consists of the separation of oil, water, and solids in two stages. The first stage comprises an API separator, a corrugated plate interceptor, or other type of oil/water separator. Figure 1–5 shows an API separator.

An API separator is designed to allow the wastewater to flow slowly through the unit, allowing sufficient time for free oil to float to the surface and be skimmed off, while solids settle to the bottom of the unit and are removed by mechanical scraping and depositing of the sludge into a collection hopper.

The second stage relies on physical or chemical methods to separate emulsified oils from the wastewater. Physical methods are

Figure 1–5 An API separator in operation.

typically a series of settling ponds designed to have long retention times to allow settling, or they use dissolved air flotation (DAF). DAF technology can range from simple circular basins equipped with a configuration of submersible nozzles, which bubble pressurized air up through a near stagnant pool of wastewater, to dynamic systems designed with multiple compartments, which capitalize on different water retention times in each compartment and a combination of settling, flotation, and impingement to remove contaminants. In the simplest DAF systems, air is bubbled through the wastewater, and both the oil and suspended solids are skimmed off of the top.

Chemical pretreatment is used to aid in the treatment process. Chemicals such as ferric hydroxide or aluminum hydroxide are used to coagulate impurities into a froth or sludge, which can be skimmed off the surface. The wastes from the API separator, the primary treatment sludge, and sludge derived from other gravitational separation equipment, the float from DAF units, and the wastes from settling ponds are all regulated hazardous wastes streams in the United States.

Reclaimed wastewater from primary treatment, in some cases, can be discharged to a publicly owned treatment works (POTW) for final treatment, but most facilities find the need to perform secondary treatment. One very important reason for investing a higher degree of treatment is to produce reclaimed water of sufficient quality that it can be recycled through the refinery. Water in many parts of the world is now recognized as a valuable resource and responsible companies focus on conservation and recycling practices. In some instances secondary treatment allows a sufficiently high-quality effluent that it can be discharged to a nearby surface water body. In the United States, this requires that the facility maintain a National Pollution Discharge Elimination System (NPDES) permit.

In secondary treatment, the dissolved oil and other organic pollutants often are consumed biologically, using carefully controlled microorganisms. Biological treatment generally requires the addition of oxygen through a variety of techniques, including activated sludge units, trickling filters, and rotating biological contactors. Secondary treatment stages generate biomass waste, which can be further treated under anaerobic conditions and dewatered.

A third stage of treatment is known as *polishing*. Polishing is used when a higher degree of reclaimed water quality is desired and most often is needed to meet legal discharge limits. Polishing steps involve such technologies as granulated activated carbon (GAC), rapid sand filtration, filtration, and other methods aimed at removing biomass, silt, trace metals, various inorganic chemicals, and any remaining organic species.

Figures 1–6 through 1–13 show typical operational units in a refinery setting. These photographs show the more-common wastewater treatment plant operations, most of which can be found throughout primary treatment stages. Some wastewater streams require separate and special treatment technologies. Generally, these technologies are applied as pretreatment operations for certain streams, which then are sent to be combined with the main wastewater flows in the treatment works. As an example, the sour water drained from the distillation reflux drums cannot be readily handled in a primary treatment stage.

Figure 1–6 Close up of the API separator with highly turbid, oily water.

Figure 1–7 A flocculation unit for chemical pretreatment in operation.

Figure 1–8 A circular air flotation unit in operation.

Figure 1–9 A small modular DAF in operation.

Figure 1–10 The water collection system following the air flotation unit.

Figure 1–11 A simple aeration pond designed to enhance microbial activity in the polishing stages of treatment.

Figure 1–12 A stabilization pond used in the primary section of treatment. Both chemical addition and pH control are applied.

Figure 1–13 A settling pond in operation. Sedimentation is used in conjunction with the DAF. It may be used in either or both primary and secondary treatment stages, depending on the influent wastewater characteristics and desired effluent quality.

Sour water contains dissolved hydrogen sulfide, various organic sulfur compounds, and ammonia. These are best handled by the use of an absorption tower, where they may be stripped out of solution then sent to the main treatment plant.

Wastewater treatment plant operations themselves are a major source of air emissions that require control. Air releases are generally fugitive emissions and obnoxious odors from numerous vessels, ponds, contactors, and sewer system drains. The treatment plant also generates large quantities of solid waste or sludge, much of which are regulated hazardous materials. These require special handling and posttreatment operations to stabilize, contain, and transport them off-site for disposal.

1.5.2 Sour Gas Treatment and Sulfur Recovery

Sulfur is removed from several refinery off-gas process streams to meet SO_x emission standards under the Clean Air Act (CAA). Process off-gas streams, known as *sour gas*, are derived from the coker, the catalytic cracking unit, hydrotreating units, and the hydroprocessing units. These streams contain high concentrations of lethal hydrogen sulfide mixed with light refinery fuel gases. Simple technology involves stripping the SO_x and H_2S, as shown by the example in Figure 1–14. However, stripping alone simply transforms an air pollution problem into a water pollution problem. Since elemental sulfur is a saleable product that can be used in acid and fertilizer manufacturing, its economical recovery makes sense for refinery operations.

Figure 1–14 A hydrogen sulfide stripper in operation at a refinery.

Before elemental sulfur can be recovered, the fuel gases (mainly methane and ethane) need to be separated from the hydrogen sulfide gas stream. This can be accomplished by dissolving the hydrogen sulfide in a solvent. Common solvents used are amines, such as diethanolamine (DEA). Also, dry adsorbents such as molecular sieves, activated carbon, iron sponge, and zinc oxide are effective. In an amine solvent absorption tower, gases are contacted with a DEA solution or another appropriate amine solvent, and the H_2S is dissolved in the solution. The fuel gases are removed for use as fuel in process furnaces in other parts of the refinery. The amine-hydrogen sulfide solution then is heated and steam stripped to separate out the hydrogen sulfide.

Current methods of recovering sulfur from H_2S gas streams are typically done by a combination of two processes: the Claus process, which can be followed by a tail-gas treatment unit (TGTU) for very high sulfur recovery efficiencies. Since the Claus process by itself cannot achieve a high removal efficiency, a TGTU is used to further recover additional sulfur to obtain an overall sulfur removal efficiency of 99.7%.

The Claus sulfur recovery process is conventional technology used to remove sulfur-based compounds from fossil fuels. Its use can be found in the sulfur recovery portion of most refinery operations. Automatic, continuous, and accurate monitoring of H_2S and SO_2 are critical factors in the optimization of the Claus process. The main reactions of interest during the Claus sulfur recovery process are

Reaction Furnace: $3H_2S + \frac{3}{2}O_2 \rightarrow SO_2 + H_2O + 2H_2S$
Catalytic Converter: $2H_2S + SO_2 \rightarrow 2H_2O + 3S_v$

For efficient conversion it is important to control the stoichiometry in the sulfur kiln. An $H_2S{:}SO_2$ ratio of 2:1 is required for efficient recovery. As such, the process is optimized when the air demand is zero. The air demand can be used as a feedback control parameter and is defined as $2[SO_2] - [H_2S]$. The tail gas contains a number of sulfuric compounds, including H_2S, SO_2, COS, CS_2, and sulfur vapor. For increased efficiencies and to be able to obtain an accurate measurement of both H_2S and SO_2, it is important to measure all components separately. This is most important for the H_2S measurement, as the COS and CS_2 present in the gas stream interfere with determination of H_2S and their concentrations must be taken into account. Even small concentrations of COS and CS_2can cause measurement errors on the order of 25%. For this reason, careful selection of instrumentation and monitoring systems are important considerations in the design stage.

The Claus process can receive H_2S-rich gas streams from the amine units and the sour-water stripping (SWS) system. Other components that enter the sulfur recovery unit (SRU) include ammonia (NH_3), CO_2, and to a minor extent various hydrocarbons.

The Claus process consists of partial combustion of H_2S-rich gas and reacting the resulting SO_2 and unburned H_2S in the presence of an activated alumina catalyst to produce elemental sulfur. Elemental sulfur is produced according to the following reaction:

$$2H_2S + SO_2 \rightarrow 2H_2O + 3S$$

The Claus process is in public domain and applied at virtually all refineries. A two-stage Claus process is the most common in Europe. There are many licensors of this process. The primary ones are as follows.

United States companies:

- Parsons
- TPA
- Black & Veatch
- Jacobs
- KBR

Other companies:

- Lurgi
- KTI (Technip)

More than a dozen processes for TGTU have been developed to enhance the recovery of sulfur compounds from refinery streams:

- Wet sulfuric acid (WSA) process, developed by Haldor Topsoe
- Direct sulfur recovery process, developed by Research Triangle Institute
- Bio-FGB process, developed by Paques
- Parson hi-activity process
- Amoco's cold-bed absorption (CBA) process
- Delta Hudson (now Jacobs) process (MCRC™)
- French Institute of Petroleum (IFP) tail gas treatment (Clauspol)
- Shell Claus off-gas treatment (SCOT)
- Lo-Cat® process, licensed by Merichem
- Marathon Oil's hysulf process
- Recycle selectrox process, developed by Parsons and Unocal and licensed through Universal Oil Process (UOP)

Other processes are

- SULFREEN™
- HYDROSULFREEN™
- SUPERCLAUS™
- CANSOLV®
- CLINTOX
- Z-SORB

Not all of these are commercial at the time of this publication. Only a few of these are briefly described next.

The Wet Sulfuric Acid Process

The wet sulfuric acid (WSA) process was developed by Haldor Topsøe A/S to produce 98% sulfuric acid. It can be used as a TGTU, but if higher levels of sulfur recovery are desired, this technology is limited.

The Direct Sulfur Recovery Process

The direct sulfur recovery process (DSRP) is under development by the Research Triangle Institute (RTI) to treat the regeneration off-gas from metal oxide sorbent-based coal gas desulfurization processes. This process is currently in the pilot plant testing stage. It has not been commercialized.

The Shell-Paques Process

The Shell-Paques process, originally developed by Paques, is a biological process for removal of H_2S from gas streams. In this process, a gas stream containing H_2S contacts an aqueous soda solution containing sulfur bacteria in an absorber. The soda absorbs the H_2S, and the solution goes to a regenerator. The regenerator consists of an aerated atmospheric tank where H_2S is biologically converted to elemental sulfur. Sulfur may be recovered as a moist filter cake or pure liquid sulfur.

This process is in the early stages of commercialization, and no applications are in tail gas claus unit (TGCU) service at the present time.

The Hi-Activity Process

The hi-activity process was developed by Parsons. It is a direct oxidation process similar to the SUPERCLAUS™ process, which involves using air to convert H_2S directly to sulfur over a special catalyst according to the following reaction:

$$H_2S + \frac{1}{2}O_2 \rightarrow S + H_2O$$

It is not expected that this process can guarantee a 99.7% sulfur recovery, but indications are it can come close. Two units have been built at the time this book was written.

The Cold-Bed Absorption Process, MCRC™, and SULFREEN™

The cold-bed absorption (CBA) process, MCRC™, and SULFREEN™ processes are the same, only developed by different companies. These processes are a dry-bed, subdewpoint, adsorption process based on the extension of the modified Claus process to low temperatures. The process uses activated alumina catalyst in a reactor operating below the sulfur dewpoint (127–149°C) to form elemental sulfur. The Claus conversion is increased because of the greater equilibrium conversion at subdewpoint temperatures and because the sulfur is adsorbed on the catalyst, which further shifts the reaction to the right according to the following reaction:

$$2H_2S + SO_2 \rightarrow 3S + 2H_2O$$

However, when the sulfur load on the catalyst exceeds a certain quantity, the catalyst activity decreases. To restore catalyst activity to an acceptable level, the catalyst must be periodically thermally regenerated and the sulfur recovered. This regeneration step is accomplished by flowing hot process gas, usually the first reactor effluent, through the deactivated bed. After sulfur desorption, the catalyst is cooled to the operating temperature.

This process typically has a sulfur recovery in the range of high 98% to low 99%. Numerous units are in operation.

Clauspol

The Clauspol process and its variations were developed by the French Institute of Petroleum (IFP). The process is based on extending the Claus reaction under subdewpoint conditions in a liquid phase. The liquid phase is a nonvolatile solvent in which H_2S and SO_2 are soluble but not liquid sulfur.

The SCOT Process

The SCOT (Shell Claus off-gas treating) process, developed by Shell, is the most commercially proven tail-gas process. The basic process consists of heating and reducing all sulfur species to H_2S, cooling and quenching, and absorbing the H_2S in an amine solution then recycling it back to the front of the Claus unit.

A SCOT unit is capable of recovering 99.99% of the sulfur entering the SRU.

Lo-Cat®

The Lo-Cat® process is currently owned by Merichem. The process uses a chelated iron solution that reacts with the H_2S and forms sulfur. The chelated iron then is regenerated with air. The Claus must be operated at a high H_2S/SO_2, since SO_2 is not converted to sulfur.

The process is capable of recovering as much as 99% of the inlet sulfur. However, the chemical costs of the process are several hundred dollars per ton of sulfur recovered. Only one Lo-Cat® is being used as a TGTU.

The Hysulf Process

The Hysulf process was developed by Marathon. This process has two steps. H_2S reacts with t-butyl anthraquinone (tBAQ) dissolved in NMP (N-methyl-2-pyrolidone) in one step. The NMP functions as an agent to transfer hydrogen from the H_2S to the quinine:

$$NMP + H_2S \rightarrow [NMPH]^+ + H^-$$

The complex then reacts with the anthraquinone to yield sulfur and hydroquinone:

$$t\text{-}BaQ + [NHPH]^+ + HS^- \leftrightarrow t\text{-}BAHQ + NMP + \tfrac{1}{8}S_8$$

The solid S_8 is centrifuged or filtered from the NMP solvent. The reduced hydroquinone is reoxided back to the quinine by catalytic dehydrogenation in a second step. The NMP-hydroquinone stream is heated to dehydrogenation temperature (274°C) and passed through a catalytic reactor at atmospheric pressure. The hydroquinone dehydrogenates and hydrogen is liberated:

$$t\text{-}BAHQ \rightarrow t\text{-}BAQ + H_2$$

Hysulf was never commercialized by Marathon, which stopped all development on this process.

The Selectrox Process

The Selectrox was originally developed by Union Oil of California. The process consists of hydrogenating the sulfur species (SO_2, COS, CS_2, S_v) back to H_2S. This stream is cooled, quenched, and air is added to the stream in the presence of the Selectrox catalyst, where a portion of the H_2S is oxidized to SO_2. The H_2S and the SO_2 then react in the classic Claus reaction to form sulfur.

The expected recoveries are in the low 99%. This process is complex, and very few are in operation.

HYDROSULFREEN™

The SULFREEN™ process was developed by TotalFinaElf and Lurgi Oel-Gas-Chemie. A limitation of the SULFREEN™ and all the subdewpoint processes is that it does not convert COS or CS_2. These tail-gas constituents typically account for 50% of the sulfur losses from a SULFREEN™ unit. HYDROSULFREEN™ was developed to reduce this problem. Effectively, a fourth-stage Claus reactor with preheat and a sulfur condenser is added to achieve the HYDROSULFREEN™ configuration. The reactor employs a catalyst, such as CRS31, that by hydrolysis converts most of the COS and CS_2 to H_2S as shown by the following reactions:

$$COS + H_2O \rightarrow H_2S + CO_2$$
$$CS_2 + 2H_2O \rightarrow 2H_2S + CO_2$$

The reactor operates at about 300°C at its inlet. Air is added to maintain H_2S/SO_2 ratio at 2 in the subdewpoint (SDP) reactors. Downstream of the condenser, the tail gas goes to two conventional SDP reactors.

This process typically has a sulfur recovery in the range of low to mid 99%. A limited number of units are in operation.

SUPERCLAUS™

The SUPERCLAUS™ process was developed by Comprimo, now Jacobs Comprimo Nederland. The SUPERCLAUS™ process is a modification of the Claus process involving a specially designed catalyst for selective oxidation of H_2S to sulfur. While SUPERCLAUS™ is not a TGTU in the sense that a SCOT unit is, it is proven to increase overall sulfur recoveries to as high as 99+%. The conventional Claus reaction is reversible as shown in the following reaction:

$$2H_2S + SO_2 \rightarrow 3/n\ S_n + 2H_2O$$

The SUPERCLAUS™ catalyst promotes the following reaction:

$$2H_2S + O_2 \rightarrow 2/n\ S_n + 2H_2O$$

The SUPERCLAUS™ catalyst does not promote the reverse Claus reaction of

$$3/n\ S_n + 2H_2O \rightarrow 2H_2S + SO_2$$

Numerous units are in operation.

CANSOLV®

CANSOLV® was originally developed by Union Carbide. It is now marketed by Cansolv Technologies Inc. CANSOLV® is a new process proposed for treating tail gas based on burning the tail gas to convert the sulfur species to SO_2, which is absorbed in the company's proprietary solvent (chemical solvent) and the SO_2 is stripped from the solvent and returned to the Claus plant feed.

CANSOLV® should be able to guarantee recoveries in excess of 99.7%; however, no operating CANSOLV® units are in TGTU service.

CLINTOX

The CLINTOX process was developed by LINDE AG. It is one of the newer processes proposed for treating tail gas based on burning the tail gas to convert the sulfur species to SO_2, which is absorbed in the company's solvent (physical solvent). The SO_2 is stripped from the solvent and returned to the Claus plant feed.

CLINTOX should be able to guarantee recoveries in excess of 99.7%; however, no operating CLINTOX units are in TGTU service at the present time.

Z-SORB

Z-SORB was developed by Phillips Petroleum Co. The Z-SORB process is based on a proprietary regenerable zinc oxide absorbent. The H_2S in the tail gas is absorbed and converted to zinc sulfide by the bed. When the bed is loaded, it is regenerated using a diluent air stream. The zinc sulfide is converted back to zinc oxide and the sulfur goes to SO_2. The regeneration gas stream containing the SO_2 then is recycled to the Claus unit. The Claus unit is operated to produce a high H_2S/SO_2 ratio, since the zinc oxide does not react with SO_2. Z-SORB should be able to guarantee recoveries in the mid 99%; however, Phillips is no longer offering this process for license.

1.5.3 Additive Production and Hydrogen Production Technologies

Several chemicals are added to motor fuels to meet improved performance criteria or federal and state environmental requirements. These chemicals are mostly alcohols and ethers. Since about the 1970s, alcohols and ethers have been added to gasoline to increase octane levels and reduce carbon monoxide generation in replacement of lead additives. Lead additives were phased out under the 1970 CAA.

In 1990, a more-stringent CAA amendments established minimum and maximum quantities of chemically combined oxygen in motor fuels and an upper limit to vapor pressure. The consequence of this is that alcohol additives increasingly have been supplemented or replaced by a number of different ethers that can better meet the oxygen requirements and vapor pressure limits.

Common ethers employed as additives are methyl tertiary butyl ether (MTBE) and tertiary amyl methyl ether (TAME). Larger refineries manufacture their own stocks of MTBE and TAME by reacting isobutylene or isoamylene with methanol. Smaller refineries purchase their feedstock from either larger refineries or chemical suppliers.

Isobutylene is usually obtained from several refinery sources, including light-end naphtha from the FCCU and coking units, the byproduct from steam cracking of naphtha and light hydrocarbons during the production of ethylene and propylene, catalytic dehydrogenation of isobutene, and the conversion of tertiary butyl alcohol recovered as a byproduct in the manufacture of propylene oxides. Several technology routes to MTBE and TAME from isobutylene and methanol are possible. Many of these use a two-stage acidic ion exchange resin catalyst system. The reaction is exothermic, and cooling to within a narrow range is critical to achieving optimal conversion efficiency. The process produces MTBE or TAME along with a small stream of unreacted hydrocarbons and methanol. Methanol is extracted in a water wash. The resulting methanol-water mix is distilled to recover the methanol for recycling.

Methanol (also ethanol) can be produced from biomass, coal, and coke (petcoke) by a variety of technologies and used for on-board reforming for transportation purposes. These technology routes are also sources of hydrogen. Refineries are always seeking new sources of hydrogen as well as new markets. The majority of these technologies have been developed as coal gasification technologies and applied in power plant applications, but commercialization routes driven by various market incentives have steered developers toward biomass and other raw feedstocks. The primary configurations are

- Entrained flow
- Fluidized bed
- Moving bed

Typical operating characteristics of coal gasifiers are shown in Table 1–1. The following text provides brief descriptions of the key features of the major technologies.

Entrained Flow Technologies

Hitachi

This coal gasification technology is based on an oxygen-blown entrained flow gasifier, where the majority of experience has been gained in a 150 t coal/d unit. The gasifier is a water-cooled tube lined by a high-temperature-resistant castable. Pulverized coal is

Table 1–1 Typical Operating Characteristics of Coal Gasifiers

	Moving Bed	*Fluidized Bed*	*Entrained Flow*
Exit gas temp., °C	420–650	920–1050	1200
Coal feed size	<50 mm	<6 mm	<100 mesh
Ash conditions	Dry/slagging	Dry/agglomerating	Slagging

pneumatically transported by nitrogen to the gasifier where it is injected into the gasifier chamber through two types of burners at a pressure of 2.5 MPa. The two sets of burners are installed tangentially to the gasifier sidewall, allowing a spiral flow of coal and oxygen from the upper stage to the lower stage and making particle residence times much longer than those of a gas stream. Enough oxygen is fed to the lower burner to melt the slag. Molten slag solidifies on the gasifier wall as a first layer and subsequent molten slag flows over the layer of the solidified slag to the slag tap hole at the bottom of the gasifier, where it is quenched with water and finally removed via a lock hopper. Coal fed to the upper burners is reacted at lower temperature with a smaller amount of oxygen; it is then gasified and converted to reactive char. The char moves down along the spiral gas flow and mixes with high-temperature gas in the lower portion of the gasifier, where gasification proceeds further. The raw gas produced, together with the fly ash and the remaining char particles, go up toward the exit of the gasifier. They enter a syngas cooler, where they are cooled to 450°C prior to going through a cyclone and a filter that retain most of the fly ash and the char particles, which are finally reinjected into the gasifier by pneumatic transport under nitrogen. The syngas goes successively through a water scrubber to remove halides and is desulfurized to be cleaned enough to comply with the strict tolerance limits of fuel cells.

SCGP

The Shell coal gasification process (SCGP) can operate on a wide variety of feedstocks. It consists of three principal stages:

1. Gasification (partial oxidation), in which the feedstock is converted to syngas in the presence of oxygen and a moderating agent (steam) in a refractory-lined gasification reactor.
2. Syngas effluent cooler (SEC), in which high-pressure steam is generated from the hot syngas leaving the reactor.
3. Carbon removal, in which residual carbon and ash are removed from the syngas in a two-stage water scrubbing unit.

The Shell gasifier is a dry-feed, pressurized, entrained slagging gasifier. Feed coal is pulverized and dried with the same type of equipment used for conventional pulverized coal boilers. The coal is then pressurized in lock hoppers and fed into the gasifier with a transport gas by dense-phase conveying. The transport gas is usually nitrogen; however, product gas can be used for synthesis gas chemical applications, where nitrogen in the product gas is undesirable. The oxidant is preheated to minimize oxygen consumption and mixed with steam as a moderator prior to feeding to the burner. The coal reacts with oxygen at temperatures in excess of 2500°F to produce principally hydrogen and carbon monoxide with little carbon dioxide. Operation at elevated temperatures eliminates the production of hydrocarbon gases and liquids in the product gas. The high-temperature gasification process converts the ash into molten slag, which runs down the refractory-lined water wall of the gasifier into a water bath, where it solidifies and is removed through a lock hopper as a slurry in water. Some of the molten slag collects on the cooled walls of the gasifier to form a solidified protective coating. The crude raw gas leaving the gasifier at 2500–3000°F contains a small quantity of unburned carbon and about half of the molten ash. To make the ash nonsticky, the hot gas leaving the reactor is partially cooled by quenching with cooled recycle product gas. Further cooling takes place in the waste heat recovery (syngas cooler) unit, which consists of radiant, superheating, convection, and economizing sections, where high-pressure superheated steam is generated before particle removal.

The first commercial IGCC (integrated gasification combined cycle) plant using the Shell coal gasification process is in Buggenum in the Netherlands, for which the construction was completed in 1993. The plant achieves an overall efficiency of 43%, which could be increased to over 50% if using the most-recent gas turbines. The Buggenum design processes coal with natural gas as backup. The plant can process up to 2000 t/d of fuel. A demonstration plant (220 t/d) at the Oil Deer Park Manufacturing complex in Houston completed tests proving the ability of the SCGP to gasify more diverse types of coals (220 t/d of bituminous coals or 365 t/d of high-moisture, high-ash lignite) before being shut down in 1991. Any coal that can be milled to the right size and pneumatically transported can be gasified in the Shell-entrained flow gasifier. Some adjustments have to be made to keep the SCGP performances optimal when changing coal. Bituminous coals, in most cases, require steam injection and oxygen/MAF (moisture- and ash-free) coal ratios from 0.85 to 1.05 for producing a syngas with a CO/H_2 ratio of 2.2–2.4 and 1–2.5% CO_2. Subbituminous coals and lignites normally do not require steam injection and can be operated with an oxygen/MAT coal ratio between 0.8 and 0.9, producing syngas with some 3–5% CO_2 and a CO/H_2 ratio of 2.0–2.2. Anthracites require a higher oxygen/MAF coal ratio of 1.0–1.1, a higher steam/oxygen ratio of 0.15–0.3, and produce a syngas with similar CO_2 contents to bituminous coal (1–2.5% CO_2, but a higher CO/H_2 ratio of 2.4–2.6). The ash content of a coal has an impact on the performance of the SCGP process in terms of efficiency, as slag forms part of the insulation of the wall of the gasifier and prevents excessive heat loss during the gasification reaction.

Sulcis, a new IGCC project based on the SCGP technology was being developed in Sardinia, Italy. The project is presently in standby due to financial reasons but expected to go ahead. It was planned to have similar characteristics as the Buggenum plant. The Sulcis plant has been designed to gasify 5000 t/d blends of local coal (high-sulfur, high-ash subbituminous coal) and imported LHV (low-heating value) coals.

A large IGCC demonstration plant is also planned to be built at the Yantai Power plant in Shandong province in China. Technical prefeasibility studies were carried out in 1994–95. Development prospects were predicted and comparisons made with CFBC (circulating fluid-bed combustion), PFBC-CC (pulsed fluid-bed combustion, combined cycle), and supercritical units. Two 400 MW IGCC units should be installed. Their net efficiency is planned to be more than 43%. They are designed to gasify bituminous coals with high sulfur content (2.5–3%) from Yanzhou in Shangong. Sulfur will be recovered as elemental sulfur with a predicted removal efficiency of 98%. Three other gasification plants are planned to be developed by Shell in partnership with Sinopec in China and a fourth one is under feasibility study. All the plants will produce syngas for ammonia/urea production or H_2 for other chemical plants (methanol, oxo), replacing naphtha reformers, oil gasifiers, or outdated coal gasifiers.

MHI

Mitsubishi Heavy Industries (MHI) consists of an air-blown MHI gasifier divided into two sections: a lower combustion section connected by a diffuser to an upper reducing section. Dry pulverized coal is fed at two points into the gasifier with half the coal being fed into the combustor together with air where it is burned to produce CO. The temperature inside the combustor is sufficiently high to melt the coal ash without the addition of flux. The slag runs to the bottom of the gasifier, where it is quenched in a water bath and removed using a lock hopper system. The gas produced in the combustor rises to the reducing section, where the remaining coal is added. Coal then is gasified in the reducing section to produce a low CV (caloric value) syngas mainly formed of nitrogen. As the reducer section is at a lower temperature than the combustor section, any molten ash carried upward is solidified. The syngas produced exits the gasifier through a syngas cooler, then cyclones are used to collect the chars as the coal is not completely gasified in the reducing section. Chars collected in the cyclones are reinjected at the base of the gasifier to ensure complete carbon conversion. Because of the very high temperatures reached in the combustion section, this type of gasifier is well suited to gasify the very-high-ash-melting-point Australian coals with no addition of a fluxing agent. The MHI gasification technology has been tested in Nakoso (Japan) in two pilot-scale gasifiers. A new IGCC project has been started: a 250 MW air-blown IGCC demonstration plant located in Nakoso, where the former pilot plants were based and will process up to 1500 t/d of coal, which is about nine times more than in the former 200 t/d pilot plant. The system will have a unique feature: The oxidizing gas will be partially extracted from the gas turbine compressor and enriched with oxygen coming from an independent air separation unit, making the gasifier operation more stable and giving a certain flexibility to the system that does not exist in the two highly integrated European IGCC plants. An advantage of the MHI two-stage dry-fed entrained flow gasifier, compared with the one-stage gasifiers, is that the syngas temperature at the outlet of the gasifier is not as high as the one flowing out of a one-stage gasifier. This means that the process does not require a large radiant cooler or a quenching system to mix cold recycled gas with the syngas. The overall cost of the process should be less than that of existing IGCC plants. The raw gas produced, together with the fly ash and the remaining char particles, go up toward the exit of the gasifier. They enter a syngas cooler, where they are cooled to 450°C prior to going through a cyclone and a filter that retain most of the fly ash and the char particles, which are finally reinjected into the gasifier by pneumatic transport under nitrogen. The syngas goes successively through a water scrubber to remove halides and is desulfurized to be cleaned sufficiently to comply with the strict tolerance limits of fuel cells.

Texaco

The gasifier is a pressure vessel with a refractory lining that operates at temperatures in the range 1250–1450°C and pressures of 3 MPa for power generation and up to 6–8 MPa for H_2 and chemical synthesis. The feedstocks, oxygen, and steam are introduced through burners at the top of the gasifier. Solid feedstocks, such as coal, are preprocessed into a slurry by fine grinding and the addition of water. The slurry is pumped into the burner and the water, which is added with the slurry, replaces most of the steam that should normally be injected into the system. Raw gas and molten ash produced during coal gasification flow out toward the bottom of the gasifier. Two alternatives are available for the recovery of the ash and cooling the raw gas. The raw gas can be either cooled and cleaned from the slag ash by water quenching or cooled in a radiant syngas cooler from 1400°C to 700°C. The heat recovered in the second option then is used to raise steam to be used in the process or to generate power. Molten slag flows down the heat recovery steam generator and is quenched in a batch at the bottom of the cooler and finally removed through a lock hopper system. The quench alternative is the preferred option for coal feedstocks, as they could contain traces of salts (sodium and calcium) that could be corrosive for the syngas coolers at high temperatures. However, this alternative could be a disadvantage for power generation, as thermal efficiency is slightly lowered.

Several existing projects are based on Texaco technology and among them an IGCC project, the Polk power station managed by Tampa Electric Corp. During the first three commercial years of operation, 10 coals or coal blends were tested to determine the cheapest feedstock to process while respecting new environmental regulations. The slag removal system of the Polk power station was designed for processing coals with a maximum of 12% (Wt, dry basis) ash content. The operating temperature of the gasifiers has to be high enough for the coal mineral matter to melt and flow freely down the bottom of the gasifier. Texaco fixed the minimum heating value of the coals at 30 MJ/kg to produce enough syngas to fully load the combustion turbine. It would be necessary to increase the oxygen supply size as well as the slurry delivery system capacity to run the plant with a lower-heating-value coal. The plant is designed to accommodate coals with sulfur contents of up to 3.5% (Wt, dry basis). Expensive modifications of the acid gas removal system were required for higher-sulfur-content coals than the first base coal (Pittsburgh No. 8-1 with a sulfur content of around 2.5%). Following major problems, the company decided to switch to coal blends with a lower sulfur content. The limit in chlorine concentration in the coals was fixed at 0.15% (dry ash). A higher concentration of chlorine in coals would damage the system. Other coal properties have an influence on the technical and economic aspects of the Texaco-based IGCC operation and necessitate coal testing in the device prior to selecting them for the Polk power station.

The Texaco technology also is used for chemical plants. Five chemical plants of this type were built after 1993. In the United States, Eastman Chemicals (Kingsport, Tennessee) owns two Texaco quench gasifiers that operate at 70 bar and 1400°C for the production of acetic acid and acetic anhydride. Although the facility is configured to make acetyl chemicals, the company claims that gasification and cleanup plants are completely compatible with an electric power option, and in fact, an electric power option of 523 MWe is reported to be under development at Kingsport. This is in line with the new projects of cogeneration of chemicals and electricity sponsored by the U.S. Department of Energy (DOE) under the Vision 21 program. Another U.S. company, Waste Management & Processors, Inc., is presently conducting a techno-economic feasibility study in partnership with Texaco, Sasol, and Nexant for the development of one of the three demonstration early entrance coproduction plants (EECP) under the Vision 21 program. The objective is the commercialization of a coal gasification/liquefaction technology to produce ultraclean Fisher-Tropsch transportation fuels with either power, chemicals, or steam as coproducts. The proposed plant location is at the Gilberton Power Plant cogeneration facility, in Pennsylvania. It involves the gasification of local waste coals, mainly high-ash-content anthracite wastes, derived from an on-site coal cleaning operation that contains coal fines, coal dust, and dirt. Another demonstration EECP project is being developed by Texaco in collaboration with Rentech (Fisher-Tropsch Technology), Brown and Root services, Praxair, and GE Power Systems for the production of electricity and chemicals from coal and petroleum coke. The project involves technical and economic studies of several process options, including syngas composition, Fisher-Tropsch product upgrading, wastewater treatment, catalyst/wax separation, acid-gas removal, tail-gas utilization, and site selection. There is also a plan for the construction by coal power of a 430 MW IGCC plant based on the Texaco technology near the Hatfield colliery in the north of England. The IGCC project with CO_2 removal and production of H_2 is being studied by Jacobs consultancy in cooperation with GE. The IGCC power plant is configured to be capable of removing 75% of the feed carbon as CO_2 prior to combustion in the gas turbine. By performing a "sour shift" of the syngas, most of the carbon monoxide should be converted into carbon dioxide and an equal volume of hydrogen. If carbon dioxide removal is performed, then the fuel for the combustion turbine will consist mainly of H_2.

BBP

Babcock Borsig Power (Noell; BBP) technology, also known as the *Noell entrained flow technology*, was first developed in 1975 in the former East Germany for the gasification of lignite in a 3 MW pilot plant. A full-scale (130 MW) gasifier was built in the 1980s to produce syngas and town gas. The technology was known as the GSP process before being acquired by Noell in 1991. The process can be a dry-fed or liquid-fed oxygen-blown slagging gasifier. If solid fuel is to be gasified, it is first pulverized, then pneumatically conveyed to the feeding system and dry fed together with oxygen and steam through a burner located at the top of the gasifier. Depending on the fuel ash content, the gasification chamber can be covered by either a cooling screen or a cooling wall. Both the refractory and the solid slag provide thermal insulation and maintain the tube surface temperature below 230°C. To allow the solidified slag to regenerate continuously, only fuels with an ash content of more than 1 wt %, such as coals, can be processed in the gasifier lined with a cooling screen. Heat removed by the cooled tube wall represents 2–3% of the total heat produced during gasification and is used to generate low-pressure steam. Syngas saturated with water is further cooled to 150–200°C and recycled to the quench sprays within the gasifier. The bottom part of the gasifier consists of a quench bath to cool and solidify the slag, which is then removed in a granular form. The only Noell gasifiers in commercial operation are at Schwarze Pumpe (Germany) and a newly built one, the BASF Seal Sands, located in Middlesbrough in the United Kingdom. The BBP research and development center, based at Freiberg (Germany), comprises two facilities with capacities of 5 and 10 MW. The smaller one was originally designed in 1979 for the gasification of both solid and pulverized solid materials. The pilot plant is presently used by Dow Chemical for the development of the gasification of chlorinated wastes. The second one was also designed for the gasification of pulverized materials (coal, waste), liquids, and slurries (waste oil, sludge, paint waste) and built in 1997. A wide range of coals from anthracites to brown coals have been gasified in the two pilot plants since the 1980s. BBP claims it is capable of providing appropriate test conditions to optimize feedstock preparation prior to gasification as well as to determine the optimum gasification conditions for more than 80 fuels including 30 coals.

E-GAS

The E-GAS (formerly Destec) coal gasifier is a slurry-feed, pressurized, upflow, entrained two-stage slagging gasifier. The dry coal concentrations in the slurry range from 50 to 70 wt %, depending on the inherent moisture and quality of the feed. Part of the coal slurry (80%) is injected with oxygen (95%) through two burners at the lower stage of the gasifier, where it is partially combusted at a temperature of 1350–1400°C and a pressure of 3 MPa. Molten ash formed flows down the gasifier and is removed through a taphole into a water quench. There is no lock hopper for ash removal, which has the advantage of reducing the overall height of the system. The fuel gas produced in the lower stage flows upward in the upper stage, where it can react with the remaining 20% coal slurry. This two-stage process presents the advantage of producing a gas with a higher calorific value than the one produced in a one-stage process. The crude gas exiting the gasifier at a temperature of around 1050°C is cooled to 370°C in a fire-tube syngas cooler. This unit generates saturated high-pressure steam. The fire-tube syngas cooler is a boiler system with the hot gas circulating on the boiler side as opposed to a water syngas cooler in which water circulates in tubes in a syngas tank. The fire tube reportedly is considerably cheaper than the ones used in the Shell, Texaco, and Prenflo processes. After the cooling step, the syngas is cleaned with filters to remove large ash and char particles, which are pneumatically reinjected into the gasifier. The filter elements, made of metal for an acceptable resistance to corrosive syngas, are periodically backpulsed with high-pressure syngas to remove particulate cake formed on their surface. The particulate cake falls to the bottom of the vessel and is pneumatically transferred, together with the high-pressure syngas, to the first stage of the gasifier, where it is recycled. Finally, the particulate-free syngas proceeds to the low temperature heat recovery system, where it is scrubbed with sour water condensed from the syngas to remove troublesome chlorides and trace elements that could cause corrosion within the piping and vessels as well as form undesirable products in the acid-gas removal system. After scrubbing and reheating, the syngas enters the COS hydrolysis unit where the COS present in the syngas is converted to H_2S.

The syngas then is cooled through a series of shell and tube exchangers to 35°C before entering the acid-gas removal system. This cooling step also condenses water from the syngas. Most of the ammonia (NH_3) and some of the carbon dioxide (CO_2) and H_2S present in the syngas are absorbed in the water as dissolved gases.

Wabash River (Indiana) plant is the only E-Gas gasifier in operation. Prior to the repowering of the Wabash River plant to an IGCC, some tests of bituminous coals, including high-sulfur coals, were performed in a 2200 t coal/d plant based in Plaquemine, Louisiana, in the early 1990s. The Wabash River power plant is designed to use a range of local coals with a maximum sulfur content of up to 5.9% (dry basis) and a higher heating value of 31.4 MJ/Kg (moisture and ash free). It is presently operating on Illinois No. 6 coal. Alternative fuels (petcoke) also have been tested successfully at Wabash River, and future tests may include coal fines. Coal fines are believed to be a promising fuel in the locality of the Wabash River facility, as they are produced by the operations of the adjacent mine. They are also available from surface reserves, where the fines have been landfilled in the past and are predicted to be 40–60% cheaper than the present coal delivered to the facility.

Prenflo

In the Prenflo facility, coal is fed together with oxygen and steam through four burners located at the lower part of the gasifier. Syngas is produced at a temperature of 1600°C and quenched at the gasifier outlet with recycled cleaned gas to reduce its temperature to 800°C. Then, the syngas flows up a central distributor pipe and down through evaporator stages before exiting the gasifier at a temperature of 380°C. The raw gas next is dedusted in two ceramic candle filters and a part of it is recirculated into the syngas cooler. The syngas is finally washed in a Venturi scrubber. Slag formed during the gasification process is quenched in a water bath then removed through a lock hopper system. The only commercial-scale unit is based in Puertollano, Spain (capacity of 338 MWe). It is the largest unit worldwide based on solid fuels. The plant has been operating since 1996 and can process up to 2600 t/d of coal/petcoke fuel mixed with limestone (2% weight) and produces 180,000 m^3/d of raw gas. The annual production of slag (85% of the ash in weight) and fly-ash (15% of the ash in weight) are, respectively, 120,000 t and 12,000 t per year. The demonstration project has attained commercial development with a gross efficiency of 47.2% (net efficiency of 42%).

Fluidized-Bed Technologies

HTW

The HTW (high-temperature Winkler) process was first developed by Rheinbraun in Germany to gasify lignites to produce a reducing gas for iron ore. The gasifier consists of a refractory-lined pressure vessel equipped with a water jacket. Feedstocks are pressurized in a lock hopper located below the coal storage bin and pneumatically conveyed to a coal bin. The conveying gas is filtered and recirculated. Coal in the receiving bin then is dropped via a gravity pipe into the fluidized bed, which is formed by particles of ash, semi-coke, and coal. The gasifier is fluidized from the bottom with either air or oxygen/steam and the temperature of the bed is kept at around 800°C, below the fuel-ash-fusion temperature. An additional gasification agent is introduced at the freeboard to decompose, at higher temperature (900–950°C), undesirable byproducts formed during gasification. The operating pressure can vary from 1 to 3 MPa, depending on the use of the syngas. The raw syngas produced is passed through a cyclone to remove particulates and cooled. Solids recovered in the cyclones are reinjected into the gasifier, and dry ash is removed at the bottom via a discharge screw.

The syngas cooling system has been the subject of study as to whether to use a water-cooled or fire-tube syngas cooler. The main reason was that the existing water-cooled syngas cooler faced fouling and corrosion problems. A conventional water scrubber system was originally used for gas cleaning, but due to blockages, fouling, corrosion, and the high operating cost of the system, Rheinbraun decided to develop a hot gas filtration system. A hot gas ceramic candle unit formed of 450 candles was developed and operated for 15,000 hours. The HTW technology manufactured by Rheinbraun was successfully applied for the synthesis of chemicals (methanol) from lignites at Berrenrath (Germany) between 1986 and 1997. The plant was shut down at the end of 1997, as at the time, the process was no longer considered economically viable. A similar commercial plant has been operating in Finland since 1988, essentially with peat for the production of ammonia. A 140 t coal/d pressurized HTW gasification plant was also commissioned and built at Wesseling (Germany) in 1989, to supplement the research and development of the HTW technology for coal use and particularly to study its future application to an IGCC process for power generation. The plant was designed for a maximum thermal capacity of 36 MW and operated for three years either as an air-blown or an oxygen-blown gasification plant with pressures up to 2.5 MPa. A wide range of coals was tested in the Wesseling plant, including brown coals and a highly volatile bituminous coal (Pittsburgh No. 8). The Wesseling plant provided the operational data required to design a potential 300 MW commercial IGCC power plant (KoBra), which was finally never built. However, there is a project to develop a 400 MW IGCC plant based on the HTW technology (two units) to replace 26 existing Lurgi moving beds at Vresova in the Czech Republic. The new HTW plant (80 t/h coal and pressures up to 3 MPa) should operate on Czech lignite and will benefit from years of research and development at the Wesseling and Berrenrath plants. To adapt the HTW technology to the Czech lignites and also to the pre-existing Vresova IGCC plant (coal grinding plant, air separation unit, waste water treatment, and steam turbine), tests were performed by Rheinbraun in a HTW bench-scale gasification unit and compared to results obtained with other coals in the same bench-scale unit and in a demonstration plant.

IDGCC

The IDGCC (integrated drying gasification combined cycle) technology was specifically developed for the gasification of high-moisture, low-rank coals by Herman Research Pty Limited in Morwell, Australia. The gasifier is a 5 MW air-blown pressurized fluidized-bed pilot plant fed with coal from an integrated drying process. The feed coal is pressurized in a lock hopper system and fed into the dryer, where it is mixed with the hot gas leaving the gasifier. The heat in the gas is used to dry the coal while the evaporation of water from the coal cools the gas without the need of expensive heat exchangers. The gasifier operates at 900°C under 2.5 MPa air pressure. Chars and ash are collected at the bottom of the gasifier and from a ceramic filter and burned in a separate boiler. The final ash product is similar to that from a conventional low-rank boiler. A wide range of low rank coals could be processed in the IDGCC, with only small changes in the operating conditions. Coals containing high levels of sulfur can be processed with sorbents, such as limestone or dolomite, directly injected into the bed. This would obviate the need for additional cooling of the gas to 40°C for sulfur removal from the very high moisture syngas. The extra cooling would have led to a very large energy loss from water condensation and reduced mass energy for the gas turbine. It is expected that the IDGCC could handle coals with lower moisture content and higher ash content. As the IDGCC plant is based on a fluidized-bed gasification technology, it is then not recommended, as in most of the fluidized-bed technologies, for coals with relatively low reactivities or low ash melting points. When looking at environmental considerations and particularly at the concept of CO_2 removal and H_2 production, the IDGCC, which produces a very moist syngas, can provide the water for the shift reaction with no or much reduced robbing of the steam cycle and may have potential for future development. It was reported that the IDGCC process is more efficient and, as a consequence, more environmentally friendly (lower CO_2 emission) than conventional processes and would be just slightly less efficient than an Australian black coal IGCC process.

KRW

Coal and limestone, crushed to below ¼", are transferred from feed storage to the KRW fluidized-bed gasifier via a lock hopper system. Gasification takes place by mixing steam and air (or oxygen) with the coal at a high temperature. The fuel and oxidant enter the bottom of the gasifier through concentric high-velocity jets, which assure thorough mixing of the fuel and oxidant and of the bed of char and limestone that collect in the gasifier. On entering the gasifier, the coal immediately releases its volatile matter, which burns rapidly, supplying the endothermic heat of reaction for gasification. The combusted volatiles form a series of large bubbles that rise up the center of the gasifier, causing the char and sorbent in the bed to move down the sides of the reactor and back into the central jet. The recycling of solids cools the jet and efficiently transfers heat to the bed material. Steam, which enters with the oxidant and through a multiplicity of jets in the conical section of the reactor, reacts with the char in the bed, converting it to fuel gas. At the same time, the limestone sorbent, which has been calcined to CaO, reacts with H_2S released from the coal during gasification, forming CaS. As the char reacts, the particles become enriched in ash. Repeated recycling of the ash-rich particles through the hot flame of the jet melts the low-melting components of the ash, causing the ash particles to stick together. These particles cool when they return to the bed, and this agglomeration permits the efficient conversion of even small particles of coal in the feed. The velocity of gases in the reactor is selected to maintain most of the particles in the bed. The smaller particles carried out of the gasifier are recaptured in a high-efficiency cyclone and returned to the conical section of the gasifier, where they again pass again through the jet flame. Eventually, most of the smaller particles agglomerate as they become richer in ash and gravitate to the bottom of the gasifier. Since the ash and spent sorbent particles are substantially denser than the coal feed, they settle to the bottom of the gasifier, where they are cooled by a counter-flowing stream of recycled gas. This both cools and classifies the material, sending lighter particles containing char back up into the gasifier jet. The char, ash, and spent sorbent from the bottom of the gasifier flow to the fluid-bed sulfator, where both char and calcium sulfide are oxidized. The CaS forms $CaSO_4$, which is chemically inert and can be disposed of in a landfill. Most of the spent sorbent from the gasifier contains unreacted CaO. Sulfur released from burning residual char in the sulfator is also converted to $CaSO_4$.

Pinon Pine in Nevada is the only large-scale coal-based IGCC plant (100 MWe) using the KRW technology, and it is also the only one designed with a 100% hot-gas cleanup. The demonstration plant, owned by Sierra Pacific Resources and sponsored by the U.S. DOE has had numerous problems. The gasifier had 18 startups, and all of them failed due to equipment design. Successes in the project included operation of the combined cycle portion of the plant at 98% availability, efficient removal by the hot-gas filter of particulates from the syngas, and production of a good-quality syngas for only 30 hours since the first syngas was produced in 1998. Sierra Pacific Resources, which owns the Pinon Pine power plant, was going to be sold to WPS Power Development but the sale has been suspended by the state of Nevada, which placed a moratorium on the sale of power plants in the state.

ABGC

The ABGC (air-blown gasification cycle) is a hybrid system developed at pilot scale (0.5 t/h coal capacity) by the former Coal Technology Development Division of British Coal. The gasifier is based on a spouted-bed design and operated at pressures up to 2.5 MPa and a temperature between 900 and 1000°C. Coal fed in the gasifier produces a gas with a low CV of around 3.6 MJ/m^3. Sorbents, such as limestone, are injected into the gasifier to retain up to 95% of the sulfur originally present in coal. Syngas is first cleaned in a cyclone, then cooled to around 400°C, and cleaned by a ceramic filter finally to be burned and expanded through a gas turbine. Only 70–80% of the fuel is gasified, and partially gasified char and other solid residues (fly-ash and sulfided sorbent residues) produced in the gasifier are then transferred to an atmospheric-pressure circulating fluidized-bed combustor (CFBC) operating at a temperature of about 1000°C. Heat generated by the combustion of the char supplies a steam cycle used to drive a steam turbine to supplement the electricity generation. The ABGC process is forecast to have an efficiency of about 46–48%. The ABGC technology was later purchased by Mitsui Babcock Energy Limited (MBEL), which in collaboration with GEC Alsthom and Scotish Power PLC produced a design of a demonstration plant while being supported by the European Commission under the THERMIE program. A wide range of U.K. coals and international steam coals were studied for use in the ABGC. A laboratory at the Imperial College of Science Technology and Medicine in London studied the impact of several coal characteristics on the gasification reactivity of some internationally traded coals in bench scale reactors that could mimic the behavior of single coal particles in the ABGC. Coal characteristics studied included coal maceral composition and coal mineral matter composition.

BHEL

The BHEL is a 168 t/d coal capacity air-blown pressurized fluidized-bed gasifier IGCC pilot plant (6.2 MWe) built at Hyderabad, India, following previous gasification tests in a 18 t/d coal capacity IGCC fluidized-bed gasifier pilot plant and in a 150 t/d coal moving bed IGCC pilot plant. The plant consists of a refractory-lined reactor with a 1.4 m inside diameter in the bed, expanding to a 2 m inside diameter at the upper section of the gasifier. Crushed coal (6 mm or less in size) is injected into the system via a lock hopper and a rotary coal feeder and pneumatically transported into the gasifier with a portion of the air used by the plant. The dry granular ash produced during gasification is withdrawn from the bottom of the gasifier through a water-cooled screw extractor and discharged periodically through an ash lock system. Three refractory cyclones operating in series are used for primary gas cleaning. Fines collected in the first two cyclones can be recycled in the gasifier, but the cyclone fines, without recycling, could also be collected through a lock hopper. The gasifier operates at a temperature of 1000°C and pressure of 1.3 MPa to generate a coal gas with a net calorific value of 9.8 MJ/kg. The 168 t/d coal demonstration plant was commissioned in 1996 and has since undergone a series of tests in stand-alone and IGCC mode, operating for a total of 1200 hours by the year 2000. The plant is designed for the gasification of Indian coals with a high ash content of up to 42%.

Transport Reactor

The Kellogg transport gasifier is a circulating-bed reactor concept that uses finely pulverized coal and limestone. The gasifier is currently in development, which may lead to a commercial design. The small particle size of the coal and limestone is expected to result in a high level of sulfur capture. Additionally, the small particle size will increase the throughput compared to a KRW gasifier, potentially reducing the required number of gasifier trains (or the gasifier size) and cost. The transport gasifier is conceptually envisioned as consisting of a mixing zone, a riser, cyclones, a standpipe, and a nonmechanical valve. Oxidant and steam are introduced at the bottom of the gasifier in the mixing zone. Coal and limestone are introduced in the upper section of the mixing zone. The top section of the gasifier discharges into the disengager or primary cyclone. The cyclone is connected to the standpipe, which discharges the solids at the bottom through a nonmechanical valve into the transport gasifier mixing zone at the bottom of the riser. The gasifier system operates by circulating the entrained solids up through the gasifier riser, through the cyclone, and down through the standpipe. The solids reenter the gasifier mixing zone through the nonmechanical valve. The steam and oxidant jets provide the motive force to maintain the bed in circulation and oxidize the char as it enters the gasifier mixing zone. The hot gases react with coal/char in the mixing zone and riser to produce gasification products. The gas and entrained solids leaving the primary cyclone pass through the secondary cyclone to provide final de-entrainment of the solids from the gas. The solids separated in the secondary cyclone fall through the dipleg into the standpipe. A solids purge stream is withdrawn from the standpipe to maintain a solids inventory. The gas leaving the secondary cyclone passes through a gas cooler, which reduces the gas temperature from about 1900°F to 1100°F.

Moving-Bed Technologies

BGL

The British Gas/Lurgi coal gasifier is a dry-feed, pressurized, fixed-bed, slagging gasifier. The reactor vessel is water cooled and refractory lined. Each gasifier is provided with a motor-driven coal distributor/mixer to stir and evenly distribute the incoming coal mixture. Oxygen and steam are introduced into the gasifier vessel through sidewall-mounted tuyeres (lances) at the elevation where combustion and slag formation occur. The coal mixture (coarse coal, fines, briquettes, and flux), which is introduced at the top of the gasifier via a lock hopper system, gradually descends through several process zones. Coal at the top of the bed is dried and devolatilized. The descending coal is transformed into char and passes into the gasification (reaction) zone. Below this zone, any remaining carbon is oxidized and the ash content of the coal is liquefied, forming slag. Slag is withdrawn from the slag pool by means of an opening in the hearth plate at the bottom of the gasifier vessel. The slag flows downward into a quench chamber and lock hopper in series. The pressure differential between the quench chamber and gasifier regulates the flow of slag between the two vessels. Product gas exits the gasifier at approximately 1050°F through an opening near the top of the gasifier vessel and passes into a water quench vessel and a boiler feedwater (BFW) preheater designed to lower the temperature to approximately 300°F. Entrained solids and soluble compounds mixed with the exiting liquid are sent to a gas-liquor separation unit. Soluble hydrocarbons, such as tars, oils, and naphtha are recovered from the aqueous liquor and recycled to the top of the gasifier or reinjected at the tuyeres.

Lurgi

The Lurgi dry-ash gasifier is a pressurized, dry-ash, moving-bed gasifier. Sized coal enters the top of the gasifier through a lock hopper and moves down through the bed. Steam and oxygen enter at the bottom and react with the coal as the gases move up the bed. Ash is removed at the bottom of the gasifier by a rotating grate and lock hopper. The countercurrent operation results in a temperature drop in the reactor. Temperatures in the combustion zone near the bottom of the gasifier are in the range of 2000°F, whereas gas temperatures in the drying and devolatization zone near the top are approximately 500–1000°F. The raw gas is quenched with recycle water to condense tar. A water jacket cools the gasifier vessel and generates part of the steam to the gasifier. Sufficient steam is injected to the bottom of the gasifier to keep the temperature below the melting temperature of ash.

BHEL

In a BHEL plant, the gasification media, a mixture of air and steam, is fed through a grate, which also enables ash removal. A gas cooler is used to recover part of the sensible heat of the gas produced and superheat steam for the gasifier. Further gas cooling, as well as tar condensation, are done by water quenching. Particulates are removed with a Venturi scrubber. A pilot plant has been operated for more than 5500 hours (1100 hours as IGCC) with two types of high-ash coals, Singareni coal with an ash content of 27–35% and North Karanpura coal with an ash content of 40%. The North Karanpura coal was also tested in the Lurgi pilot-scale plant at the Indian Institute of Chemical Engineering (IICT), Hyderbad, under the same gasification conditions. It resulted in a better performance of the BHEL gasifier (calorific value and cold gas efficiency) due mainly to the larger scale of the gasifier. However, the availability of the plant was affected by the poor performance of the raw gas cooler due to tar deposition and choking. A direct contact quench was subsequently designed to replace the gas cooler and overcome that problem. The performance of the moving bed gasifier also was compared to that of a pressurized fluidized-bed gasifier later developed by BHEL at the Trichy unit in Hybedarad, India. Moving-bed gasifiers produce tar-laden gas, which makes the recovery of the sensible heat of the raw gas difficult. They also need coals with a certain particle size (5–30 mm). They produce large effluents containing tars and phenolic acids requiring elaborate effluent treatment. For these reasons, BHEL decided to develop the fluidized-bed technology for processing Indian coals.

A 6.2 MWe IGCC plant was developed by BHEL at the Trichy unit in India in 1988, as part of a research program for the development of gasification of Indian coals for the production of electricity. The gasification process was based on a moving-bed technology developed in-house after experience on a Lurgi dry-ash-bed gasifier (pilot scale 24 t/d) was gained at the IICT in Hyderabad and CFRI in Dhanbad. The gasifier is a 2.7 m diameter, 14 m high-jacketed moving-bed gasifier with a coal throughput of 150 tJd. Crushed coal of 5–40 mm size with an ash content of about 35% is the design feedstock for the gasifier, which operates at 1 MPa pressure.

Different technology suppliers worldwide have developed gasifiers. The choice of the type of the gasifier is purely a factor of the fuel characteristics. Various technology suppliers for the gasification process are listed in Table 1–2.

Composition of the syngas depends on the fuel as well as on the gasification process. The typical characteristics of the syngas as generated from different fuels at some of the IGCC projects are summarized in Table 1–3.

Table 1–2 Gasification Technology Suppliers

Technology Supplier	*Coal Feed Type*	*Oxidant*	*Gasifier Type*
Texaco, USA	Water slurry	O_2	Entrained flow
Shell, USA	N_2 carrier/dry	O_2	Entrained flow
KRW, USA	Dry	Air	Fluidized bed
Lurgi, Germany	Dry	Air	Fluidized bed
British Gas/Lurgi	Dry	O_2	Moving bed
Prenflo, USA/Krupp Uhde, Germany; Deutsche-Babcock, Germany	Dry	O_2	Entrained flow
Destec Energy, USA	Water slurry	O_2	Entrained flow
IGT U-Gas, USA/ Carbona, Finland; IBIL, India	Dry	Air	Fluidized bed
Rheinbraun HTW, Germany; RWE Energie, Germany	Dry	Air	Fluidized bed
MHI, Japan; IGC, Japan	Dry	Air/O_2	Entrained flow
ABB-CE, USA	Dry	O_2	Entrained flow
VEW/Steinmuller, Germany		O_2	Entrained flow
Hitachi	Dry	O_2	Entrained flow
Noell/GSP	Dry	O_2	Entrained flow
Ahlstrom, Sweden	Dry	Air	Fluidized bed

Table 1–3 Typical Syngas Compositions

	Project						
	PSI Wabash	***Tampa Polk***	***El Dorado***	***Shell Pernis***	***Sierra Pacific***	***IBIL***	***Schwarze Pumpe***
Fuel	Coal	Coal	Petcoke/ Waste Oil	Vacuum Residue	Coal	Lignite	*
H	24.8	27.0	35.4	34.4	14.5	12.7	61.9
CO	39.5	35.6	45.0	35.1	23.5	15.3	26.2
CH_4	1.5	0.1	0.0	0.3	1.3	3.4	6.9
CO_2	9.3	12.6	17.1	30.0	5.6	11.1	2.8
N_2 + Air	2.3	6.8	2.1	0.2	49.3	46.0	1.8
H_2O	22.7	18.7	0.4	—	5.7	11.5	—
LHV, KJ/M^3	8,350	7,960	9,535	8,235	5,000	4,530	12,500
T_{fuel}, °C	300	371	121	98	538	549	38
Oxidant	O_2	O_2	O_2	O_2	Air	Air	O_2

*Lignite/oil slurry with waste plastic and waste oil.

It is likely that refineries will integrate some of the previously described technologies, especially in light of the facts that petcoke has soft markets and sources of hydrogen are a concern. These technologies will pose additional environmental management issues. The typical steps for a gas cleanup system aim at particulate removal, sulfur removal, and NO_x removal. This is achieved as follows:

- Particulate removal: Combination of cyclone filters and ceramic candle filters.
- SO_x and NO_x removal: Combination of steam/water washing and removing the sulfur compounds for recovery of sulfur as a salable product.

Hot-gas cleanup technology is currently in the demonstration phase, and the various demonstrations have not been successful so far. Wet scrubbing technology, though with a lower efficiency, still remains the preferred option for gas cleanup systems in IGCC.

1.5.4 Heat Exchangers and Boiler Operations

Refineries use an almost countless number and variety of heat exchangers for heating and cooling petroleum process streams. These devices consist of bundles of pipes, tubes, plate coils, steam coils, and other arrangements enclosing heating or cooling water, steam, or oil to transfer heat directly to or from the process stream. The bundles must be cleaned periodically to remove accumulations of scale, sludge, and oily residues.

Historically, chromium was used as a corrosion inhibitor for heat exchangers; however, because of its elimination, the wastes currently generated from cleaning heat exchangers are not a major part of the overall hazardous waste stream from a refinery's operations.

The sludge generated can contain lead and chromium; however, this is not common. Even refineries that do not produce leaded gasoline and use nonchrome corrosion inhibitors can have measurable amounts of lead and chromium that must be managed. As with all waste streams, the refinery management needs to maintain sampling and analysis programs for all waste streams to ensure that both regulatory obligations and responsible waste management practices are applied at all times. The waste streams generated from heat exchanger cleaning include oily wastewater that requires management as well.

Heat exchangers are integral parts of the boiler operations within a refinery. Boilers produce three types of steam to support refinery operations: low-, medium-, and high-pressure steam. They also generate wastewater streams. Often, significant losses can be attributed to boiler blowdown and excess condensate. As water evaporates in the boiler steam drum, solids or scale present in the feedwater are left behind. The role of blowdown is to limit the concentration of impurities in the boiler water. The right amount of blowdown is critical. Too much can result in energy losses and excessive chemical treatment cost; while too little will allow excessive concentrations of impurities to build up and damage tubes.

Watertube or drum-type boilers have two blowdown systems: continuous blowdown from the steam drum to limit the concentration of the water and intermittent blowing from the bottom headers to remove suspended solids from the boiler. The total blowdown generally is drained to the refinery's wastewater treatment plant. Minimizing this blowdown rate reduces material and energy losses. Since the temperature of the blowndown water is the same as that of the steam generated in the boiler, any savings in blowdown also reduces the makeup feedwater and the chemical treatment additives. It, therefore, is important for refineries to focus on optimum blowdown conditions to capture financial savings, limit chemical use, and conserve water.

To capture the savings, refineries need to optimize the boiler operating windows as opposed to performing random blowdown. In practical terms, this means automation. By changing from a manual system to automated controls, operator error is eliminated and system stability achieved. An automated blowdown control system[2] optimizes surface blowdown by regulating the water volume discharged in relation to the amount of dissolved solids present, conductivity, and silica or chlorides concentrations. A probe provides feedback to a controller/driving mechanism through a modulating blowdown valve. An alternative control configuration is a

[2]http://www.spiraxsarco.com/assets/uploads/PDFs/sb/P403_03.PDF.

proportional control with the blowdown rate set proportional to makeup water flow.

In addition to optimizing blowdown conditions, refineries face significant steam and condensate losses. While these are not pollution in the traditional sense, they represent material and energy losses that ultimately are financial in nature. By way of example, consider that, when condensate leaves a steam trap, it has approximately 20% of the original thermal energy contained within the steam. Returning this condensate for reuse further reduces energy consumption, eliminates the excess use of water treatment chemicals, and allows a high quality of water to be reused as waterfeed makeup. Most often the root causes for steam and condensate losses are related to malfunctioning of steam traps and the misuse of steam in ancillary operations, such as washing and cleaning (see Figure 1–15 as an example).

Often steam and condensate losses are difficult to estimate, however they represent a sizeable loss that translates into thousands of dollars. Some examples where corrective actions can lead to savings include

- Provide preventive maintenance on level controllers. Malfunction of the level controller used on a low-pressure condensate

Figure 1–15 Low-pressure steam used in a noncritical application.

tank (see Figure 1–16 as an example) means that low-pressure condensate return cannot be controlled. This will result in frequent condensate overflow.

- Identify and stop flash steam losses. As an example, flash steam open to atmosphere from a condensate collecting tank as in

(A)

(B)

Figure 1–16 (A) A condensate collection tank. (B) Returned condensate with steam leakage.

Figure 1–17 is a result of a tank being left uncovered and open for continuous natural evaporation. This observation further supports that there is a likely failure of the steam traps, since excess amounts of steam are carried along with the return condensate. A simple solution would be to cover the condensate tank and provide a vent for the normally expected steam return. By implementing a regular inspection program and preventive maintenance practices, steam and condensate losses can be dramatically reduced with subsequent financial savings.

- Apply preventive maintenance programs to steam traps, vents, and valves. Steam and other material losses often are caused by faulty equipment and poor maintenance practices. A few examples are shown in Figures 1–18 through 1–20. Minimizing these losses brings about large savings in makeup water and

Figure 1–17 Uncontrolled steam losses from an open condensate return tank.

fuel use and results in a decrease of wastewater generation and loadings to the wastewater treatment plant, as well as reduced fugitive emissions.

Preventive maintenance programs are a crucial element in maintaining heat exchangers, boilers, and other major equipment in proper working order. These programs also represent one of the important operational components of an environmental management system (EMS), as they not only affect pollution control issues but are at the core of an EMS, which is pollution prevention. In this regard, all forms of waste, whether regulated or not, represent pollution. Simply because a regulation does not address a waste issue or a particular emission does not mean that it should be ignored or given a low priority. What goes up a stack, is lost from a leaking valve or pump gland, is vented as steam, or overflows as condensate into a sewer, are not just materials—they represent dollars. And, while the losses may seem small on a local scale, when one considers the cumulative losses over time, the financial losses to an organization are enormous.

Figure 1–18 Example of steam losses that should be corrected.

Figure 1–19 Example of steam losses at boiler section.

Figure 1–20 Example of steam losses at condensate section.

1.5.5 Plantwide Blowdown Systems

A distinction is made here between boiler blowdown and general blowdown systems. The former has been discussed in relation to the utilities operations of refineries. In a general blowdown system, refinery process units and equipment are manifolded into a central collection unit or blowdown system to provide safe handling and disposal of liquids and gases. First and foremost, plantwide blowdown systems prevent overpressure. Systems are designed to either automatically vent through pressure-relief valves or allow manual drawing of materials from process units. Usually, recirculated process streams, as well as cooling water streams, are manually purged to prevent the buildup of contaminants, as explained earlier. Portions or entire contents of equipment and tankage also require periodic purging to the blowdown system prior to shutdowns, before normal turnarounds, or for emergency shutdowns.

Blowdown systems use a series of flash drums and condensers to separate out the blowdown into its vapor and liquid fractions. Liquids generally are composed of mixtures of water and hydrocarbons containing sulfides, ammonia, and various other contaminants. These are sent to the plant's wastewater treatment plant. The gaseous matter most often contains hydrocarbons, hydrogen sulfide, ammonia, mercaptans, various solvents, and a variety of other air pollutants that must be responsibly managed. Most often, these are managed through the use of a flare as shown in Figure 1–21. The primary emissions from blowdown systems are hydrocarbons when they are directly discharged to the atmosphere and sulfur oxides when they flared.

1.5.6 Blending Operations

Blending operations represent the next-to-last step in refinery processes. The operations consist of mixing products in various proportions to meet certain product specifications, like specific gravity, vapor pressure, viscosity, sulfur content, octane number, cetane index, initial boiling point, and pour point. Blending operations are conducted either in-line or in batch-blending tanks. Air emissions from blending operations are fugitive volatile organic compounds (VOCs) from the tanks, valves, pumps, and mixing operations.

Figure 1–21 A flare in operation to manage hydrogen sulfide at a refinery.

1.5.7 Storage Tanks and Loading and Unloading Operations

No refinery can operate without numerous storage tanks, such as those shown in Figures 1–22 and 1–23. Storage tanks are used to store crude oil, intermediate process feeds for cooling, for fire water, liquefied natural gas (LNG), and for finished liquid products before transporting off-site.

Figure 1–22 LNG tank farm. A motor fuel tank is shown in the background.

Figure 1–23 Finished gasoline storage tanks.

Tank bottoms are mixtures of iron rust from corrosion, sand, water, and emulsified oil and wax, which tend to accumulate at the bottom of vessels. Liquid tank bottoms are primarily water and oil emulsions and must be drawn off periodically to minimize buildup. Tank-bottom liquids and sludge also must be removed during periodic cleaning for inspection purposes. Tank bottoms can contain tetraethyl or tetramethyl lead at those refineries that still manufacture leaded gasolines. They may also contain metals and phenols. The solids (sludge) generated from leaded gasoline storage tank bottoms are hazardous wastes.

Most tanks today are equipped with floating roofs; however, even with floating tops, they account for considerable VOC emissions. Many tank losses are attributed to the failure of tank seals on gasoline storage tanks.

Loading operations for product distribution can be a source of spills as well as fugitive VOC emissions, as shown in Figure 1–24.

Figure 1–24 Gasoline tanker loading operation resulting in a spill.

1.5.8 Cooling Towers

Cooling towers are a major workhorse at any refinery operation. Figures 1–25 and 1–26 show examples of operations and common configurations. The makeup water source is used to replenish water lost to evaporation. Hot water from heat exchangers is sent to the cooling tower. The water exits the cooling tower and is sent back to the exchangers or to other units for further cooling.

Cooling towers fall into two main subdivisions: natural draft and mechanical draft. Natural draft designs rely on large concrete chimneys to introduce air through the media. Due to the tremendous size of these towers (500 ft high and 400 ft in diameter at the base) they generally are used for water flow rates above 200,000 gal/min. Usually, these types of towers are used only by utility power stations in the United States. Mechanical draft cooling towers (Figures 1–25 and 1–26) are much more widely used and common at refineries. These towers utilize large fans to force air through circulated water. The water falls downward over fill surfaces, which helps increase the

Figure 1–25 Mechanical cooling tower operation with forced draft fans.

Figure 1–26 Rectangular mechanical cooling tower in operation.

contact time between the water and the air. This helps maximizes heat transfer between the two.

Mechanical draft towers offer control of cooling rates through their fan diameter and speed of operation. These towers often contain several areas (each with its own fan) called *cells* (Figure 1–25). Heat is transferred from water drops to the surrounding air by the transfer of sensible and latent heat. Thermodynamics dictate that the heat removed from the water equals the heat absorbed by the surrounding air. The increased heat load causes the hot-water temperature to increase considerably faster than the cold-water temperature.

Cooling towers are designed to the highest geographic wet-bulb temperatures. This temperature dictates the minimum performance available by the tower. As the wet-bulb temperature decreases, so does the available cooling water temperature. The following is the summary of steps in the cooling tower design process in industry:

- The plant engineer defines the cooling water flow rate and the inlet and outlet water temperatures for the tower.
- The manufacturer designs the tower to meet this criteria on a "worst case scenario" (i.e., during the hottest months). The tower characteristic curves and the estimate are provided to the plant engineer.
- The plant engineer reviews bids and makes a selection.

Once a tower characteristic has been established between the plant engineer and the manufacturer, the manufacturer must design a tower that matches this value. The required tower size is a function of

- Cooling range
- Approach to the wet bulb temperature
- Mass flow rate of the water
- Web-bulb temperature
- Air velocity through the tower or individual tower cell
- Tower height

Water losses include evaporation, drift (water entrained in discharge vapor), and blowdown (water released to discard solids). Drift losses are estimated to be between 0.1 and 0.2% of water supply.

Even during cold-weather months, the plant engineer should maintain the design water flow rate and heat load in each cell of the cooling tower. If less water is needed due to temperature changes (i.e., the water is colder), one or more cells should be turned off to maintain the design flow in the other cells. The water in the base of the tower should be maintained between 60°F and 70°F by adjusting air volume, if necessary. The usual practice is to run the fans at half speed or turn them off during colder months to maintain this temperature range.

An improperly maintained cooling tower produces warmer cooling water, resulting in a condenser temperature 5–10°F higher than a properly maintained cooling tower. This reduces the efficiency of the chiller, wastes energy, and increases cost. The chiller consumes 2.5–3.5% more energy for each degree increase in the condenser temperature. For example, if a chiller uses $20,000 of electricity each year, it will cost an additional $500 to $700 per year for every degree increase in condenser temperature. Thus, for a 5–10°F increase, one can expect to pay $2500 to $7000 a year in additional electricity costs. In addition, a poorly maintained cooling tower has a shorter operating life, is more likely to need costly repairs, and is less reliable. The performance of a cooling tower degrades when the efficiency of the heat transfer process declines. Some common causes of this degradation include

- *Scale Deposits.* When water evaporates from the cooling tower, it leaves scale deposits on the surface of the fill from minerals dissolved in the water. Scale buildup acts as a barrier to heat transfer from the water to the air. Excessive scale buildup is a sign of water treatment problems.
- *Clogged Spray Nozzles.* Algae and sediment that collect in the water basin as well as excessive solids get into the cooling water and can clog the spray nozzles. This causes uneven water distribution over the fill, resulting in uneven airflow through the fill and a reduced heat-transfer-surface area. This problem is a sign of water treatment problems and clogged strainers.
- *Poor Airflow.* Poor airflow through the tower reduces the amount of heat transfer from the water to the air. Poor airflow can be caused by debris at the inlets or outlets of the tower or in the fill. Other causes of poor airflow are loose fan and motor mountings, poor motor and fan alignment, poor gearbox maintenance, improper fan pitch, damage to fan blades, or excessive vibration. Reduced airflow due to poor fan performance can lead ultimately to motor or fan failure.
- *Poor Pump Performance.* An indirect cooling tower uses a cooling tower pump. Proper water flow is important to achieve optimum heat transfer. Loose connections, failing bearings, cavitation, clogged strainers, excessive vibration, and nondesign operating conditions result in reduced water flow, reduced efficiency, and premature equipment failure.

The cooling tower manufacturer's operation and maintenance instructions should be followed whenever possible. Larger, more-complicated cooling towers with special filters or controls demand a comprehensive maintenance program.

Recommended References

Bendavid-Val, A., and N. P. Cheremisinoff. *Achieving Environmental Excellence: Integrating P2 and EMS for Improved Profitability.* Washington, DC: Government Institutes, 2003.

Cheremisinoff, N. P. *Handbook of Solid Waste Management and Waste Minimization Technologies.* London: Butterworth–Heinemann Publishers, 2003.

Cheremisinoff, N. P. *Handbook of Air Pollution Prevention and Control.* London: Butterworth–Heinemann Publishers, 2002.

Cheremisinoff, N. P. *Handbook of Water and Wastewater Treatment Technologies.* Butterworth–Heinemann Publishers, 2002.

Cheremisinoff, N. P., and A. Bendavid-Val. *Green Profits: A Manager's Handbook to ISO 14001 and Pollution Prevention.* London: Butterworth–Heinemann Publishers, 2001.

Cheremisinoff, N. P. *Pollution Prevention Practice Handbook.* New York and Basel: Marcel Dekker, 2001.

Environmental Design Considerations for Petroleum Refining Crude Processing Units. API Pub. No. 311. Washington, DC: American Petroleum Institute, February 1993.

Gary, J., and G. Handwerk. *Petroleum Refining—Technology and Economics,* 3rd ed. New York: Marcel Dekker, 1994.

Leeman, J. E. "Hazardous Waste Minimization: Part V, Waste Minimization in the Petroleum Industry." *JAPCA* 38, no. 6 (June 1988).

Meyers, R. A. *Handbook of Petroleum Refining Processes.* New York: McGraw-Hill Book Co., 1986.

U.S. Petroleum Refining: Meeting Requirements for Cleaner Fuels and Refineries, volumes I–VI. Washington, DC: National Petroleum Council Committee on Refining, U.S. Department of Refining, U.S. Department of Energy, August 1993.

2

Refinery Waste and Pollution Profile

Introduction

This chapter contains a summary of typical material (waste and pollution) outputs from selected refining operations. It is presented in a matrix format for easy reference. In later chapters, this information can be referred to as a starting basis for identifying environmental aspects that the facility should examine more closely for improved performance under the environmental management system (EMS).

2.1 Refinery Material Output Matrices

These matrices provide a summary of the typical air emissions, process wastewater streams, and residual solid wastes from select refinery processes.

Air Emissions

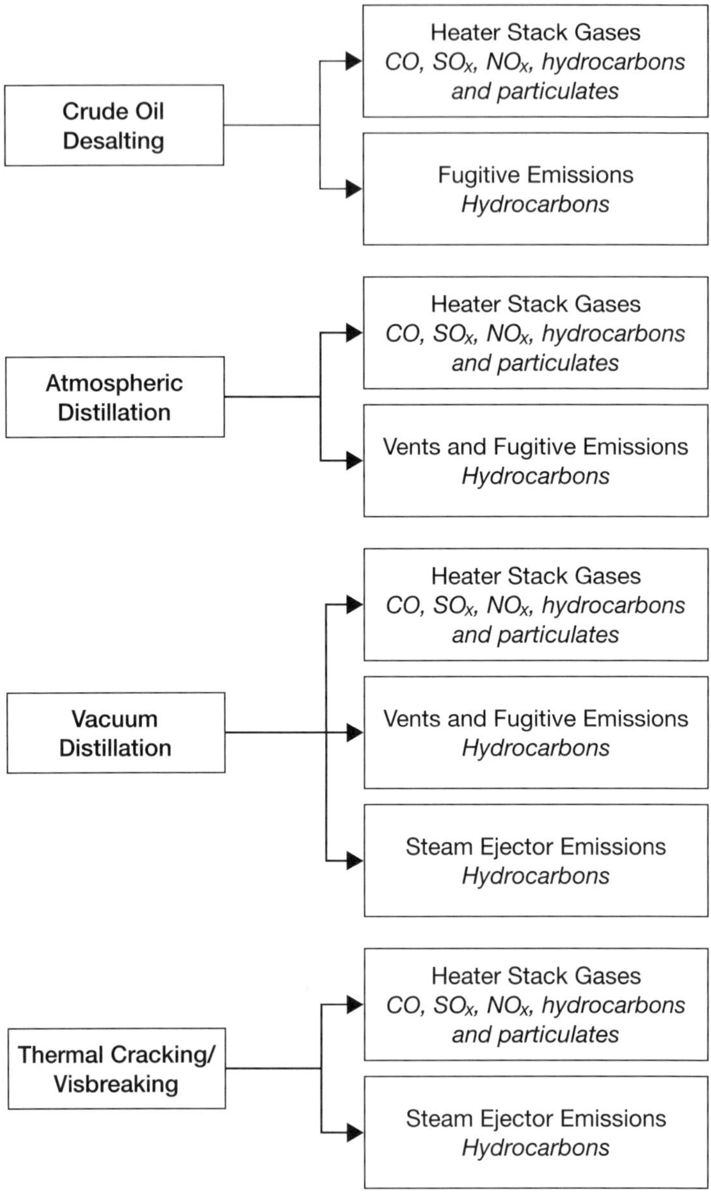

Matrix 2–1 Air emissions, part 1.

Air Emissions

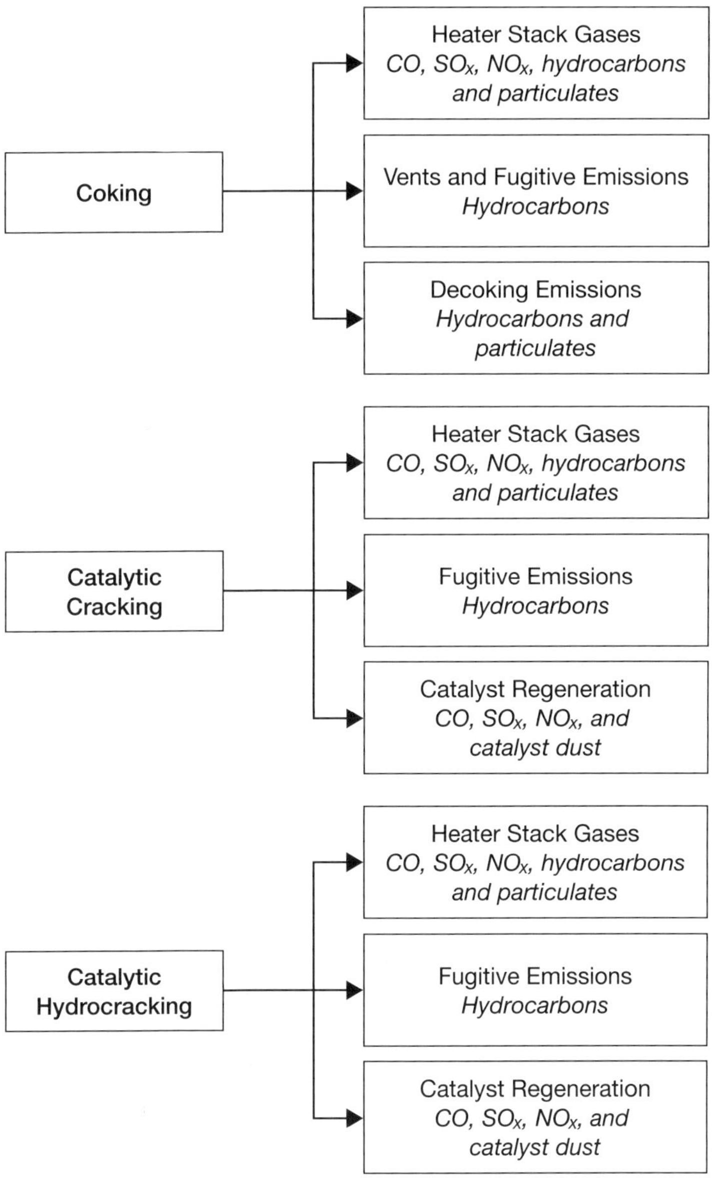

Matrix 2–2 Air emissions, part 2.

Air Emissions

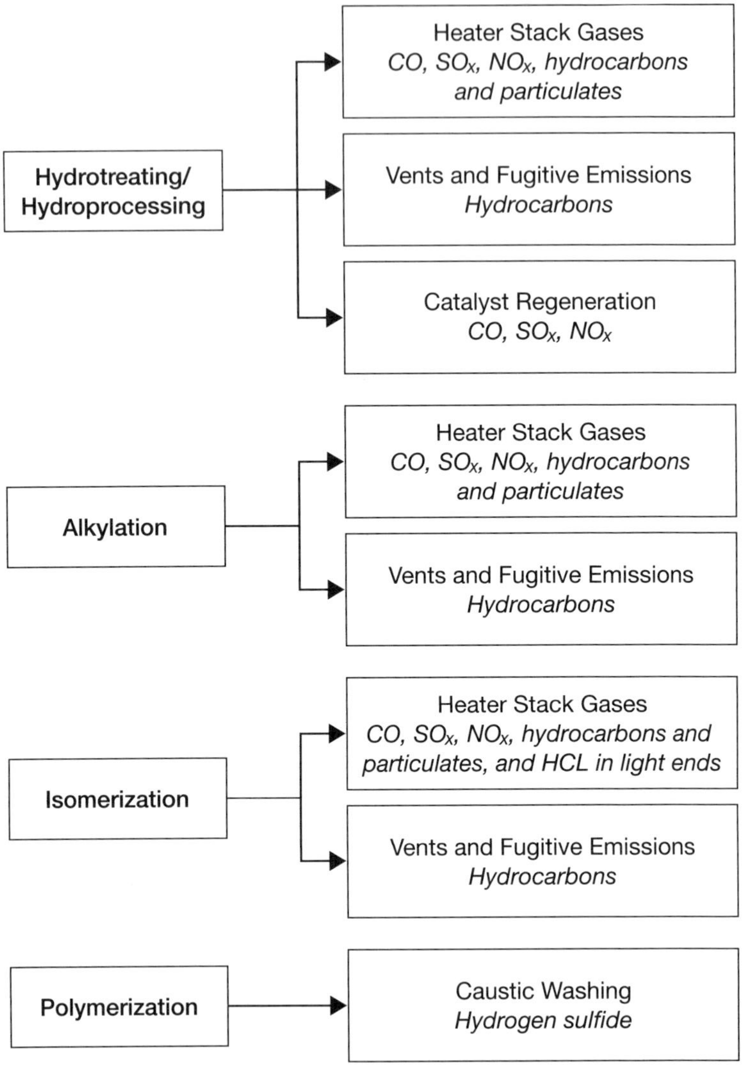

Matrix 2–3 Air emissions, part 3.

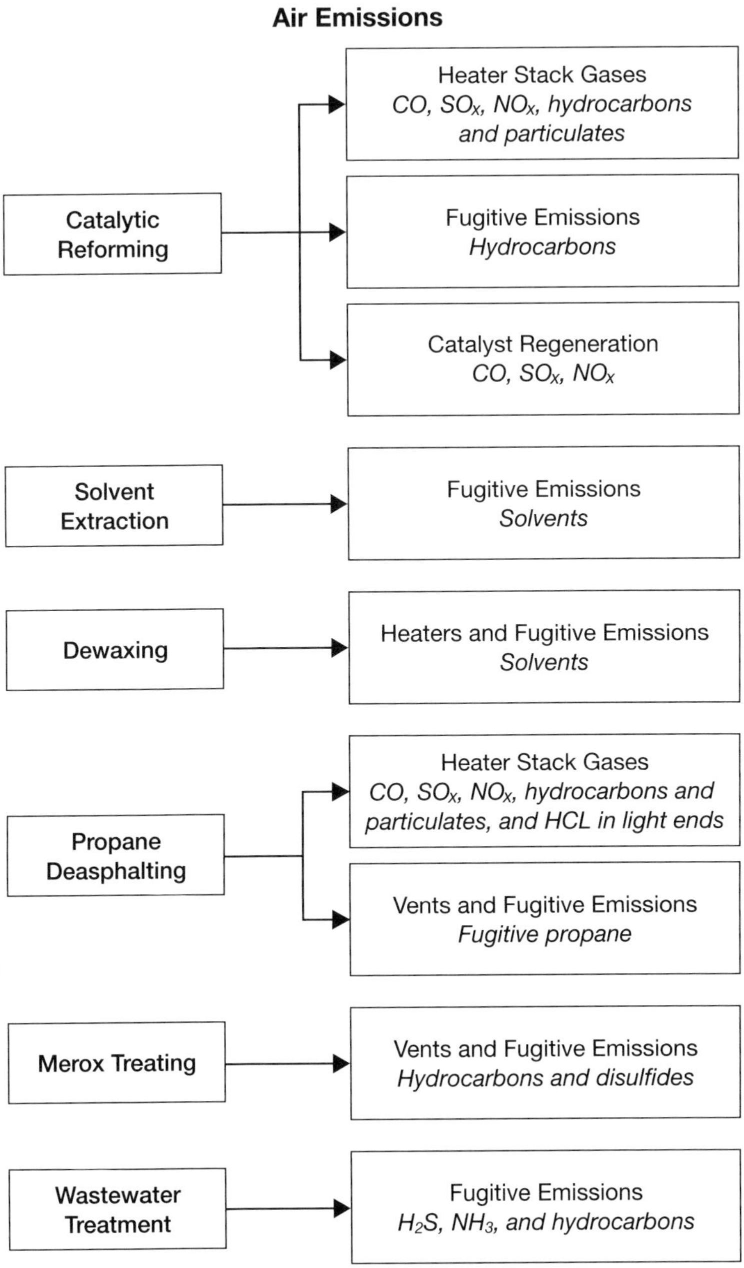

Matrix 2–4 Air emissions, part 4.

Process Wastewater

Crude Oil Desalting	Flow = 2.1 Gal/Bbl Oil, H_2S, NH_3, Phenol, High Levels of Suspended Solids, High BOD, High Temperature
Atmospheric Distillation Vacuum Distillation	Flow = 26 Gal/Bbl Oil, H_2S, NH_3, Phenol, Suspended Solids, Chlorides, Mercaptans, Phenol, High pH
Thermal Cracking/ Visbreaking	Flow = 2 Gal/Bbl Oil, H_2S, NH_3, Phenol, Suspended Solids, High BOD, COD, pH
Coking	Flow = 1 Gal/Bbl Oil, H_2S, NH_3, Phenol, Suspended Solids, High pH and COD
Catalytic Cracking	Flow = 15 Gal/Bbl Oil, High Levels of Oil, Suspended Solids, Phenols, Cyanides, H_2S, NH_3, High pH, BOD, COD
Catalytic Hydrocracking	Flow = 2 Gal/Bbl Oil, High COD, Suspended Solids, H_2S, NH_3, Low Levels of BOD
Hydrotreating/ Hydroprocessing	Flow = 1 Gal/Bbl Oil, High pH, COD, BOD, Phenol, Suspended Solids, H_2S, NH_3

Matrix 2–5 Process wastewater, part 1.

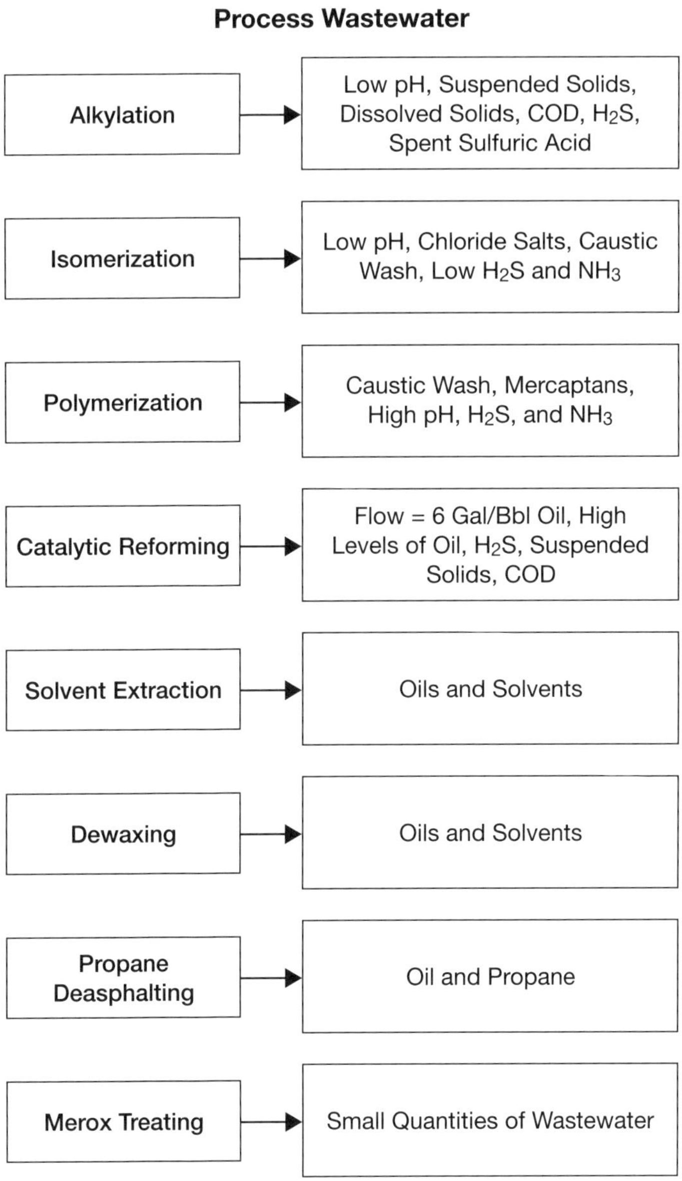

Matrix 2–6 Process wastewater, part 2.

Residual Waste Streams

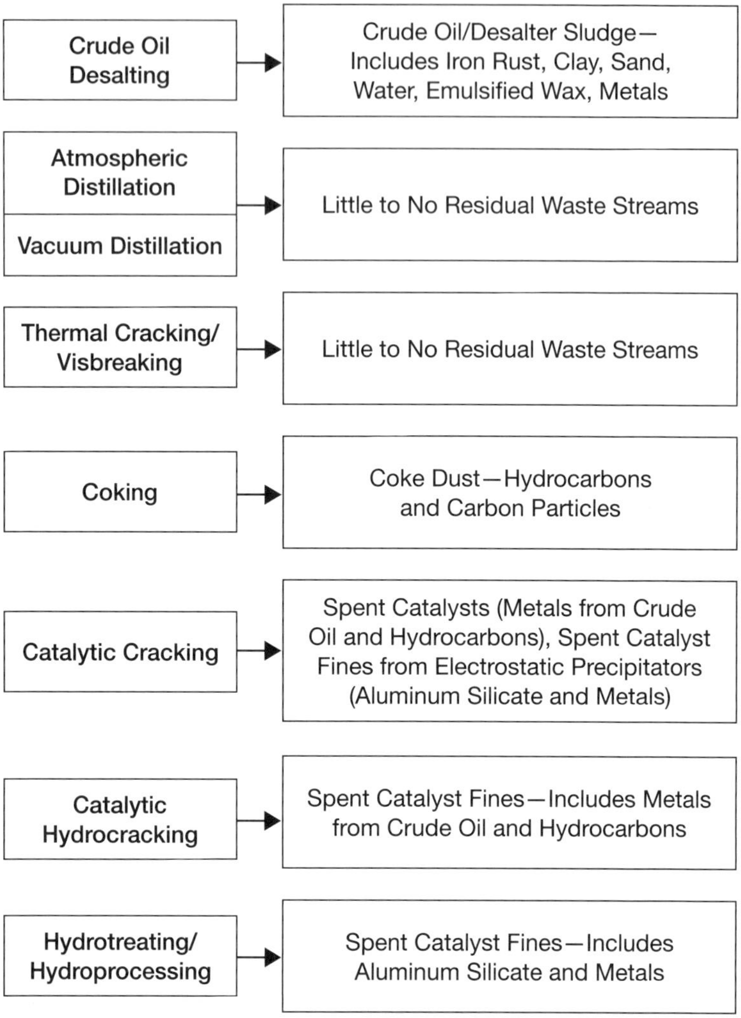

Matrix 2–7 Residual waste streams, part 1.

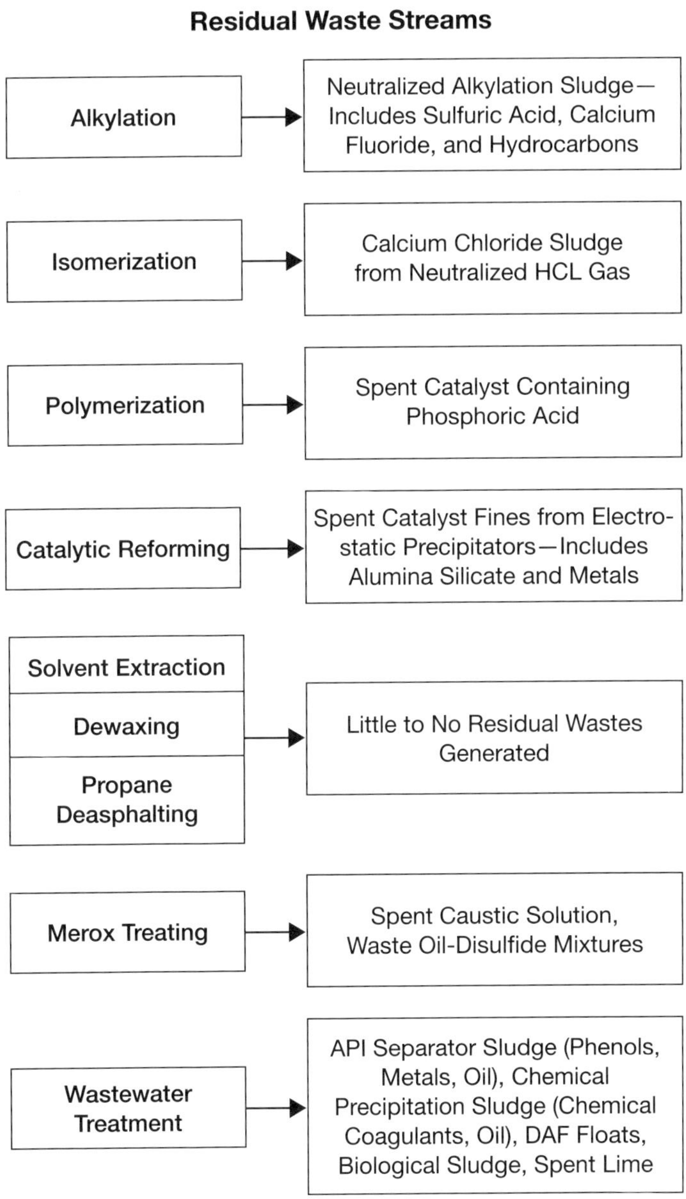

Matrix 2–8 Residual waste streams, part 2.

3

What an Environmental Management System Is All About

Introduction

This chapter provides an overview of environmental management systems. The reader is provided an explanation of what an environmental management system comprises, its intent, a description of the various systems and programs in use by businesses, and the drivers for investing into the adoption of a program. The chapter is intended both as a primer for those readers becoming acquainted with the subject for the first time and a review for those somewhat familiar with what an environmental management system is intended to do.

3.1 What an EMS Is

EMS stands for environmental management system. An EMS is a planning and implementation system that a company or other organization employs to manage the way it interacts with the natural environment. Through an EMS, a company can continually improve its environmental performance and, along with that, continually improve its economic performance and continually reduce its risk of environmental liability.

All environmental management systems are built around the way a company operates. They focus on a company's production processes and general management system. They are not intended to focus on emissions, effluents, and solid and hazardous waste, as environmental regulations do. This may sound somewhat strange for the refinery manager, because his or her intent, along with that of the environmental manager, traditionally has focused on compliance and reducing emissions to meet statutory limits. But we need to recognize that a company's emissions, effluents, and solid and hazardous waste are the result of the way a facility is managed. An EMS reduces emissions, effluents, and other wastes by improving the company's management of its processes so that less pollution and waste are generated. By doing so, compliance is achieved.

A well-designed EMS is based on proven principles of strategic management, and hence, it enables a company to address major and costly aspects of its operations proactively, strategically, and comprehensively. Without an EMS, a company can only react to environmental mishaps and disasters, to public complaints, to ever-more-stringent environmental regulations, to environmental lawsuits, and certainly to being undercut by more-progressive and -efficient competitors. A productive EMS not only improves environmental performance and reduces operating costs for a company but also reduces the risk of environmental liabilities now and into the future.

A properly implemented EMS should be integrated into the company's overall management system. Like an overall management system, an EMS is a process of continual analysis, planning, implementation, and review. It requires that top management commit itself and organize resources such as people, money, and equipment to achieve the environmental objectives of the company; and it requires that resources be committed to support the management system itself.

An EMS is like any other system of planning and implementing for continual improvement. Many continual improvement management systems, such as for product quality or worker safety, have at their core the well-known plan-do-check-act model. Under that model:

- *Plan* refers to setting company policy on the subject; systematically identifying and setting priority areas for improvement; establishing goals and targets for improved performance; and making plans to achieve those targets, especially in the highest priority areas.
- *Do* refers to carrying out those plans.
- *Check* refers to monitoring and evaluating how implementation of the plans is progressing.
- *Act* refers to revising the plans and their implementation as necessary to achieve performance targets, then moving on (some managers prefer to use the term revise or revise and act for this step).

In light of experience with this process, after a time, management would reconsider its policy, identify new priority areas for improving performance, establish new targets, devise new plans, and so on, in a never-ending process of continual improvement.

For purposes of an EMS, as the first element of "plan," top management would have to formulate an environmental policy for the company. This policy declares the company's basic environmental values and establishes guidelines that govern design and operation of the rest of the EMS. Also under "plan," a company would need to determine its compliance status regarding environmental laws and regulations, since noncompliance would represent a first-order area in which performance should be improved. After that, the company would identify and assign priority to other areas for improving environmental performance. Most often, the areas for improvement will concern air pollution, water pollution, soil contamination, solid and hazardous waste management, resource consumption, and energy and worker efficiency.

In giving these priority for the company's attention and resources, management would take into account factors such as complaints from the community, human health risks, ecological health risks, market preferences, bottom-line benefits, financing, long-term business sustainability, and risk of environmental liability in the near or even distant future.

Next under "plan," the company would establish targets for improving its environmental performance in the highest priority areas and develop one or more plans. Such plans are referred to as *environmental management programs*, *environmental action plans*, or *environmental management initiatives.*

Under "do," the company would carry out their environmental management programs or action plans. But, almost as soon as plan implementation is started, so is "check." That is, virtually from the outset of plan implementation, the company carefully monitors its environmental performance indicators of concern, and systematically tracks and evaluates implementation progress toward achieving the targets.

Under "revise" or "act," the company would modify its environmental programs in accordance with the monitoring and evaluation findings of "check." But, there is more to this element of the continual improvement model: With the hindsight of some experience, top management may now want to reconsider the company's environmental policy and even reconsider the design of its EMS.

Once the highest priority environmental performance concerns are attended to, it is time to look anew at the company's compliance needs, both now and in the future, and at other aspects of its environmental performance and set new targets, devise new environmental management programs, and so on in a process of continual improvement.

In other words, even as implementation of the original set of environmental management programs proceeds, the iterative process of continual improvement is under way. This means that even though the plan-do-check-revise/act model and most EMS models tend to be represented graphically as cycles, except for when they are put in place the first time, as a practical matter, they cannot operate effectively as a sequence of activities. Top management regularly reviews environmental policy, not only because of the output of the EMS but also because of changes in outside circumstances; a *responsible* company continuously reviews its regulatory compliance status, again, in part, because regulations are constantly changing; environmental performance is constantly monitored and new priority areas for improved environmental performance are always coming to light; and in response, new environmental action plans must be regularly developed, implemented, monitored, evaluated, and modified. Each part of the EMS operates on its own cycle, and all the parts are connected to form a continuous system. When implemented properly the environmental, economic, and risk reduction benefits of an EMS flow in a steady stream, cumulating and paying higher and higher dividends with the passage of time.

There are many types of EMSs. Some have been developed by industry associations, such as the chemical industry's *Responsible Care*, the hospitality industry's *Green Globe*, and the forest industry's *SmartWood.* Some nongovernment organizations (NGOs) have developed EMSs, such as the International Chamber of Commerce's *Business Charter for Sustainable Development.* Government agencies also develop EMSs for their own purposes. Examples are the *Facility Environmental Management and Monitoring System* of the Tobyhanna Army Depot (www.femms.com) and the North American Commission for Environmental Cooperation (NAFTA associated) *EMS Guidelines* (www.cec.org). In addition, many individual companies design their own EMSs: ExxonMobil's *OIMS* (operational integrity management system for environmental health and safety) is one example, and Baxter International's EMS based on its own adaptation of the ISO 14001 model is another. All of these in some manner follow the logic of the basic plan-do-check-revise/act model. Figure 3–1 summarizes the plan-do-check-revise/act model and the important elements of an EMS.

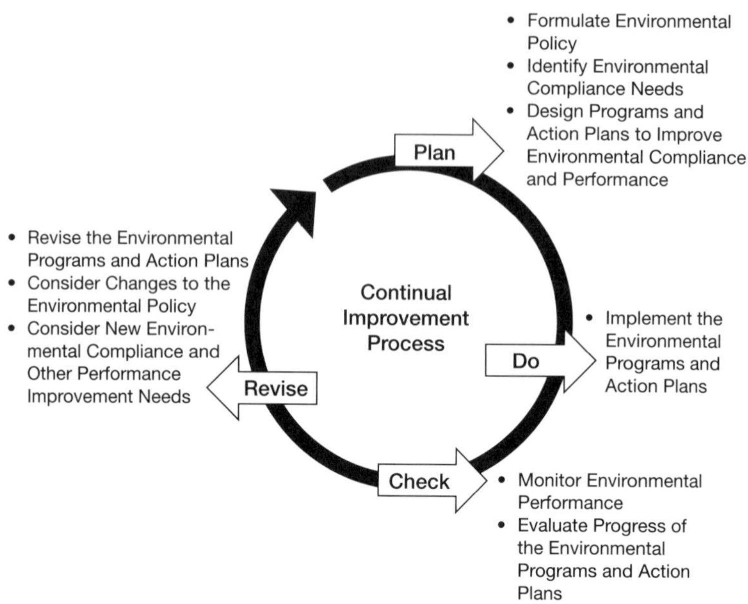

Figure 3–1 Plan-do-check-revise/act model for an EMS.

3.2 What an EMS Standard Is

3.2.1 General

An EMS is a planning and implementation system that a company or other organization employs to manage the way it interacts with the natural environment. A proliferation of different environmental management systems have been developed that are industry and special interest group specific. While each is significant and has relevance to the sector and interest group to which it applies, all are based on a standard. There are three standards on which all environmental management systems are based—BS 7750, EMAS, and ISO 14000.

3.2.2 BS 7750

BS 7750 was developed by the British Standards Institution (BSI) in 1992. It is considered to be the prototype for the ISO 14000 series and is referred to by some as the first EMS, although the roots of the chemical industry's Responsible Care program can be traced back into the late 1980s. The standard was finalized in 1996 and is also the basis for the European Union's Eco-Management and Audit Scheme (EMAS).

BS 7750 establishes standards to apply in describing a company's environmental management system, to evaluate the company's performance, and to define environmental policy, practices, objectives, and targets. It provides a catalyst for continuous improvement. The concept is similar to the use of ISO 9000 for quality systems, with the methods to be used open to definition by the company. The standard provides the framework for development and assessment of the resultant environmental management system.

BS 7750 was developed as a response to concern about environmental risks and damage (both real and potential). Compliance to the standard is voluntary for companies, and complements required compliance to statutory legislation.

BS 7750 requires an environmental policy to be in existence within the company, fully supported by top management, and outlining the policies of the company, not only to the staff but to the public. The policy needs to clarify compliance with environmental legislation that may effect the organization and stress a commitment to continuous improvement. Emphasis is placed on policy as this provides the direction for the remainder of the management system.

The preparatory review and definition of the organization's environmental effects is not part of a BS 7750 assessment; however, examination of this data provides an external audit with information on the methods adopted by the company. The preparatory review is intended to be comprehensive and takes into consideration input processes and outputs at the site. The review is designed to identify all relevant environmental aspects that may arise from existence on the site. These may relate to current operations; they may relate to future, perhaps even unplanned future, activities; and they certainly relate to the activities performed on site in the past (e.g., contamination of land or groundwater). The review also includes a wide-ranging consideration of the legislation that may affect the site, whether it is currently being complied with, and perhaps even whether copies of the legislation are available.

The company is required to declare its primary environmental aspects, which are those that can have the most significant environmental impacts. These become the primary areas of consideration within the improvement process and the company's environmental program. The program embodies a plan to achieve specific goals or targets along the route to a specific goal and describes the means to reach those objectives, so that they are real and achievable. The environmental management system thus provides further detail on the environmental program.

3.2.3 EMAS

EMAS stands for Eco-Management and Audit Scheme. This is the European Union's environmental management system standard. Its origins can be traced to the high environmental standards derived from German interests and specific companies that played a key role in its development.

EMAS was introduced by a European Union council regulation (No. 1836/93), requiring implementation in all European Union member states. The EMAS regulation was first introduced in July 1993 as an environmental policy tool devised by the European Commission, in a step toward the European Community's goal of sustainable development. The EMAS scheme was open for voluntary participation by organizations from April 1995, and its scope restricted participation to sites operating industrial activities.

In 1996, the international environmental management system standard, EN ISO 14001, was published and recognized as a step toward achieving EMAS. It was also recognized that all sectors have significant impacts on the environment and the environment benefits from good environmental management in other sectors as well.

Article 14 of the regulation allowed member states to extend the scheme to other economic sectors and several member states took the opportunity to pilot EMAS successfully in those other sectors. Article 20 of Regulation 1836/93 stated that the EMAS scheme had to be reviewed no more than five years after its entry into force.

In 2001, the new Regulation (EC) No. 761/2001 was adopted (O.J. L114, 24.4.2001, p. 1). The regulation consists of 18 articles and eight annexes. Unlike other management systems standards, the annexes form part of the regulation and are not merely informative. This means that the requirements of the annexes must be met.

The overall objective of the European Community, as defined by the Maastricht Treaty, is to "promote a harmonious and balanced development of economic activities, sustainable and non-inflationary growth respecting the environment . . . the raising of standards of living and quality of life."

The scheme does not replace existing community or national environmental legislation or technical standards nor does it remove a company's responsibility to fulfill all its legal obligations under such legislation or standards.

The goal of sustainable development calls for the use of a wider range of tools for environmental policy. The Sixth Community Environmental Action Program, Environment 2010: Our Future, Our Choice, recognizes this and aims to be a program that "completes and reinforces our body of environmental legislation where there are gaps and takes forward the implementation of our directives . . . does more in terms of mobilizing stakeholders for the environment and 'greening' the market."

Originally, the EMAS scheme was site based and open only to companies operating industrial activities. However, the EMAS scope has widened, and it now can be applied to any organization that has an impact on the environment. Participation in the scheme is open to organizations operating in all economic sectors.

EMAS identifies the different types of economic activity areas by using Statistical Classification of Economic Activity (NACE) Codes. NACE stands for *Nomenclature générale des activités économiques dans les Communautés Européennes*, which is the standard for classification of economic activities in the European Union (EU). Details of these can be found in Council Regulation No. 3037/90 (as amended by Regulation No. 761/93 and 29/2002), NACE Rev. 1. Under the previous regulation, the scheme was open to the industrial sector covering only NACE codes from 10 to 40. Now that EMAS II allows participation of organizations from all economic sectors, the whole range of NACE codes is applicable. This list enumerates NACE codes with a maximum of two decimals after the comma; however, there are NACE lists that go in much-deeper detail, with up to five decimals after the comma.

For EMAS, the NACE codes are used to classify the registered organizations and the accredited verifiers. The agreement was made that the competent body registering the site and forwarding the information to the EMAS Helpdesk should collect NACE information with a minimum of one decimal after the point. It varies from one competent body to another if they go into more detail or not.

Registration in the scheme requires an organization or company to adopt an environmental policy containing commitments both to comply with all relevant environmental legislation and achieve continuous improvements in environmental performance.

At the company or organization, an initial environmental review is undertaken. In the light of this review and the policy, an environmental program and environmental management system is established.

EMAS is prescriptive. It requires a verified initial environmental review. It requires that the policy, program, environmental management system, and details of the organization's performance are made publicly available as part of the environmental statement. EMAS is explicit in its control over contractors and suppliers, requiring that procurement issues are addressed and the company endeavors to ensure that contractors and suppliers comply with the environmental policy. EMAS specifies that organizations must attempt to "reduce environmental impacts to levels not exceeding these corresponding to economically viable application of best available technology."

To comply with the requirements of EMAS, the ISO 14001 certificate must be issued under one of the accreditation procedures recognized by the European Commission.

3.2.4 ISO 14000

ISO stands for the International Standards Organization (International Organization of Standardization). ISO is a federation of national standards bodies representing about 130 countries, headquartered in Geneva, Switzerland. It was established in 1947 as a nongovernmental organization that promotes international standardization as a way of facilitating international trade. Through an extensive collaboration process involving layers of committees and subcommittees that meet around the world, the ISO has established voluntary international standards for such diverse things as film-speed codes, the format of banking cards, freight containers, symbols for automobile controls, and screw threads.

After a successful experience developing and encouraging international standards for quality management (the ISO 9000 series), in late 1996, the ISO issued the first few standards in the environmental management series: the ISO 14000 series. ISO 14001 is the ISO standard for EMSs. Other ISO standards in the 14000 series, including some still under development, deal with subjects like auditing EMSs, environmental considerations in life cycles of products, environmental site assessments, environmental performance measures, and environmental labeling.

Within the standard ISO 14001 is the only certifiable standard, the remainder are supportive guidelines. The various standards and guidelines in the series are listed in Table 3–1.

Table 3–1 Standards and Guidelines within the ISO 14000 Standards

Series	*Subject*	*Standards*
ISO 14000	Environmental management systems	ISO 14001 ISO 14004
ISO 14010	Guidelines for environmental auditing	ISO 14010 ISO 14011 ISO 14012
ISO 14020	Environmental labels and declarations	ISO 14020 ISO 14021
ISO 14030	Evaluating environmental performance	ISO 14031
ISO 14040	Life cycle analysis	ISO 14040 ISO 14041 ISO 14042 ISO 14043
ISO 14050	Understanding terms and definitions	ISO 14050

3.2.5 Comparing the Standards

Both ISO 14000 and EMAS are based on BS 7750, and all three are similar in approach to the development of an EMS. If a company complies with BS 7750, minimal effort is needed to fulfill the requirements of either ISO 14001 or EMAS.

The European Community cites EMAS as more prescriptive than ISO 14001 and goes beyond the ISO standard. The following are some distinctions between the two:

- *Preliminary Review.* EMAS requires a verified initial environmental review, whereas ISO does not. However, ISO 14001 simply cannot be implemented without an initial environmental review. The practice when implementing ISO 14001 is to rely on external auditors. Hence, while EMAS is explicit in stating the requirement of having an external review, ISO 14001 is practically implemented in the same manner.
- *Public Availability.* EMAS requires that the policy, program, environmental management system, and details of a company's performance be made publicly available as part of the environmental statement. ISO requires only that the policy be publicly available. However, an EMS established on the basis of ISO 14001 relies on properly designed metrics aimed at tracking environmental performance. This performance is almost religiously reported by responsible companies and, in fact, is a major driver for investing in ISO 14001. Hence, the two are similar in practice.
- *Contractors and Suppliers.* EMAS is slightly more explicit in its control over contractors and suppliers, requiring that procurement issues be addressed and the company endeavors to ensure that contractors and suppliers comply with the environmental policy. ISO 14001 requires that relevant procedures are communicated to contractors and suppliers. Some companies require their preferred suppliers to adopt ISO 14001 standards. In effect, there should be no difference.
- *Commitments and Requirements.* ISO 14001 does not stipulate the extent to which performance must be improved. EMAS specifies that organizations must attempt to "reduce environ-

mental impacts to levels not exceeding those corresponding to economically viable application of best available technology."

- *Certification.* To comply with the requirements of EMAS, the ISO 14001 certificate must be issued under one of the accreditation procedures recognized by the European Commission. While certification (or registration) is the same in the United States, it is not mandatory. ISO 14001 is voluntary, and as such, many companies elect for self-declaration to minimize the initial costs for implementation. While critics argue this invalidates the EMS, the fact is, if a company truly implements an EMS, the formal registration process using independent auditors and consultants does not provide a whole lot more except cost. And, for those companies that have recently undergone ISO 9001 certification and see the value of a third-party-implemented program, costs can be kept down by utilizing the same consultants for an ISO 14001 program.

3.3 Understanding ISO 14001

3.3.1 The Basic Clauses

The clauses of the ISO 14001 EMS standard are contained in a booklet published by the ISO: *ISO 14001: Environmental Management Systems—Specification with Guidance for Use.* The reader can order a copy of the standard, learn more about it and its use worldwide, learn how the ISO 14001 certification (also called *registration*) system works, and learn more about the ISO itself at www.iso.ch. Another good website for a broad range of information on ISO 14001 for U.S. readers is www.rabnet.com.

The ISO 14001 EMS standard contains 17 brief clauses. Each clause represents an element of the EMS and contains a statement of the requirements that a company must meet to be in conformance with that element of the standard.

Clause 4.1

Clause 4.1 of ISO 14001 contains the requirement that the organization's EMS conform to the other 17 clauses.

Clause 4.2

Clause 4.2 of ISO 14001 requires that top management establish an overall environmental policy for the company in writing and requires that the policy include three fundamental commitments on the part of top management: (a) a commitment to comply with environmental regulations, (b) a commitment to try always to *prevent* pollution at the source rather than just cleaning it up after creating it, and (c) a commitment to continual improvement in environmental performance. These three commitments have become pillars of EMSs worldwide. Today, no company's environmental policy is considered credible lacking any of these three commitments. With an environmental policy like this, top management commits the company to strive for a continually improving level of environmental performance beyond what is required by law and to try to achieve it by preventing pollution.

Clause 4.3

Clause 4.3, Planning, includes assessing the current situation, both to identify aspects of the company's operations that significantly affect the environment and to inventory its compliance with environmental regulations. Then, in light of the findings of the assessment, top management, usually working through an EMS committee, establishes current environmental objectives and targets for the company. In ISO 14001, an environmental *objective* reflects the overall intent of the company regarding an environmental aspect of its operations, to reduce emissions of air pollutants by 25% per unit of production, for example. An environmental *target* is an explicit performance criterion, quantified and time framed, that represents progress toward achieving the objective, say, reducing tons of CO_2 equivalent emissions by 20% per unit of production within the first year.

Under Clause 4.3 environmental management programs are devised for achieving the objectives and targets. Implicit in this sequence of planning steps is a number of additional activities, including ranking the adverse environmental impacts of the company's operations in terms of significance (to help establish priorities for addressing the impacts) and examining alternative approaches for achieving targets.

Clause 4.4

Clause 4.4, Implementation and Operation, defines the requirements for the capabilities and supporting systems—the environmental management infrastructure—that the company must have to be in conformance with ISO 14001. This EMS infrastructure is necessary both to enable the company to carry out its environmental management programs and to control its environmental performance in the course of general operations. The ISO 14001 EMS standard calls for the following seven elements of EMS infrastructure:

- Structure and responsibility
- Training, awareness, and competence
- Communication
- Environmental management system documentation
- Document control
- Operational control
- Emergency preparedness and response

Clause 4.5

Clause 4.5, Checking and Corrective Action, contains requirements for procedures to track those characteristics of a company's operations that could contribute to significant environmental impacts. It also requires that a company monitor its performance relative to its stated environmental objectives and targets, to provide indicators of the progress and effectiveness of its environmental management programs. And it requires procedures for taking corrective action when necessary. Clause 4.5 also requires that proper measuring and monitoring records be kept, so that there is a documented history that can be reviewed and analyzed as part of the continual improvement process. Finally, this clause requires a periodic EMS audit—sometimes called an *internal audit*, *management review audit*, or *precertification audit*—as the basis for evaluating and improving the EMS itself.

ISO 14001 does not require that the company's overall environmental performance per se—the type and amount of pollution it discharges, the resources it consumes, the energy it consumes, and the like—be covered by the measuring and monitoring of Clause 4.5. But any good manager would certainly also want to monitor these things, as well the costs of pollution and waste management and cost savings associated with improved environmental performance.

Clause 4.6

Clause 4.6, Management Review, requires that top management periodically review the overall operation of the EMS and consider changes, including revision of the environmental policy, new environmental objectives and targets, new personnel assignments, and so on throughout the 17 elements of ISO 14001. The management review is central to the continual improvement process, because only top management can authorize changes in the basic structure and operation of the company's EMS.

The main body of information for this review comes from activities carried out under Clause 4.5, Checking and Corrective Action. But, changes in EMS elements may be necessary because of altered external conditions, such as environmental regulations, market preferences, supply chain requirements of a major customer, and the like; so other sources of information have to be called into play as well. Again, while ISO 14001 does not require that the management review cover the actual environmental performance of the company or the business consequences of that performance, enlightened top management also includes these subjects in the review.

ISO 14001 is an international standard for environmental management systems. It is *not* a standard for environmental performance, and it does not even suggest or imply environmental performance criteria. Conformance to the ISO 14001 standard is voluntary, and the ISO 14001 accreditation and certification mechanisms are (or should be) operated completely without government involvement. In fact, the language of the standard makes clear that it is meant to be applicable to organizations that want to self-declare their conformance with ISO 14001 or demonstrate their conformance to others (like major customers who demand it), as well as to those who seek certification or registration.

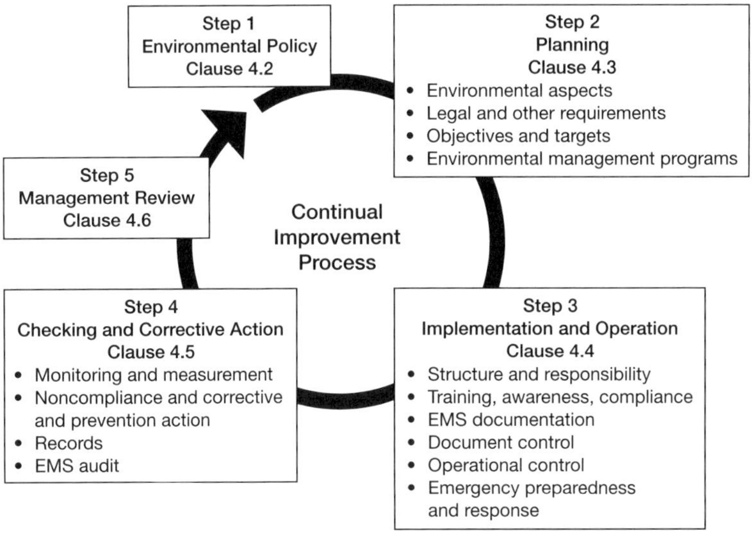

Figure 3–2 ISO 14001 within the context of the continual improvement process.

Collectively, the ISO 14001 clauses just described are based on the plan-do-check-revise/act management model. Figure 3–2 summarizes the clauses over the continual improvement process.

3.3.2 Compliance versus Conformance with Environmental Performance Standards

Compliance with regulatory environmental performance standards is a concern of governmental agencies, taken by the ISO 14001 standard as a matter apart from voluntary conformance to ISO 14001. The ISO encourages distinguishing between *compliance* with a legal requirement and *conformance* to a voluntary standard like ISO 14001. But, for a company, conformance to the ISO 14001 standard is not a matter completely apart from regulatory compliance, because if its EMS truly conforms to the requirements and spirit of ISO 14001, its environmental performance and level of regulatory compliance (not to mention its business performance) necessarily improves.

Because ISO 14001 is a voluntary standard dealing only with the requirements for an EMS, it is possible for a company to be certified as conforming to the ISO 14001 standard yet not be in compliance with environmental regulations. Stated another way, the registration or certification of ISO 14001 does not ensure that a company meets its regulatory obligations under the laws.

3.4 EMS Drivers

There are substantial costs for a company, in terms of time, staff, energy, and money, to have an EMS in place and maintain it on an ongoing basis as part of its overall management system. The justification and motivation for making such an investment stems from a large body of historical evidence in the form of case studies, testimonials, and corporate annual and environmental reports that support the idea that a well-run EMS more than pays back the investment in terms of dollar cost savings and risk reduction, not to mention other benefits of improved environmental performance and regulatory compliance.

Fundamentally, pollution is waste. Forget for the moment that environmental regulations place allowable limits on pollution. Pollution unto itself represents inputs for which money was paid that leave a manufacturing plant in a form that is not part of the company's product. This is tantamount to throwing money up a stack, down a sewer, burying it in a landfill, or into a stream. To make matters even worse from a business point of view, to meet regulatory standards companies have to pay even more money to capture the pollutants, still more money to process them, and even more money to dispose of them somewhere else, where they may become a huge financial liability some time in the future. An EMS pays for itself by providing a framework for systematically identifying and seizing opportunities to *prevent* pollution at the source, to reduce waste—through process changes, technology changes, materials substitution, or other means—and thereby improve operational efficiency and reduce the risk of financial liability. It accomplishes this by making *pollution prevention* the top commitment from a policy standpoint. That, unto itself, is enough of a driver for any competent senior manager to commit resources into implementing an EMS.

Because of the characteristics of a sound, continual improvement-based EMS and the direct economic and environmental improvements it generates, establishing and maintaining an EMS in an organization responds to a number of other drivers, which are summarized in Table 3–2.

Table 3–2 Summary of Key Drivers beyond P2 Incentives for EMS Implementation

Driver	*Explanation*
Minimize the cost of operations required for environmental management and compliance	Because an EMS is built on widely accepted management principles it integrates easily with quality management systems, occupational health and safety management systems, and the overall management system of the company, so that a minimum of added cost is required. In fact, many companies have unified environment, health, and safety (EH&S) management systems.
Improve environmental performance at lower cost	Because it builds capacity within the company for strategic environmental management, an EMS enables a company to be proactive in improving its environmental performance and compliance. Strategic environmental management includes accounting for likely future regulatory changes and continually identifying pollution prevention (P2) opportunities, an approach that always allows for lower-cost solutions than scrambling to meet new regulatory standards when they come into force.
Improve overall enterprise management and overall business performance	Because, through an EMS, environmental responsibility must be clearly allocated throughout the organization, monitoring systems must be established, adequate operational controls must be ensured, proper documentation must be created and controlled, waste and energy consumption must be minimized, and other elements of sound enterprise management must be established and maintained, an EMS promotes good business management practices overall.

continued

Table 3–2 Summary of Key Drivers beyond P2 Incentives for EMS Implementation *continued*

Driver	*Explanation*
Prepare for the coming trends in environmental regulation	There is a gradual movement throughout the world from media-based to systems-based compliance enforcement approaches that require EMS-type strategic environmental planning by top management (see, for example, EPA's National Environmental Performance Track program at www.epa.gov). A company with a productive EMS is positioned to take advantage of the many cost and other benefits of these new enforcement approaches.
Improve prospects for long-term sustainability	Because an EMS leads to continually improving environmental performance through P2 and other measures, companies that implement and maintain an EMS continually reduce their demands for energy and natural resources per unit of production. This improves a company's prospects for sustainability as resources become scarcer and more expensive. It also contributes to sustainability by reducing the risk of potentially catastrophic environmental liability. Finally, it helps assure that a company's operations will not be shut down for reasons of regulatory noncompliance.
Reduce operating costs	A well-functioning EMS continually generates opportunities to minimize waste and energy. As a result, a company that establishes and maintains an EMS always, in due course if not immediately, realizes substantial operating cost reductions. Such cost reductions—from an energy-saving improvement, for example—continue to accrue year after year. In many cases, the accrued savings increase on an annual basis over time. Increases in the costs of energy or hazardous waste disposal over the last two decades, for example, have far outpaced national inflation averages in the United States and parts of the European Community.

continued

Table 3–2 Summary of Key Drivers beyond P2 Incentives for EMS Implementation *continued*

Driver	*Explanation*
Achieve high rates of return on capital investments	An EMS systematically identifies and prioritizes process, technology, and materials substitution investment opportunities that improve environmental performance. These opportunities are usually for P2 and energy-efficiency investments yielding returns that commonly far exceed typical rates of return on other types of investments. The financial benefits of these investments can be seen to be even greater when reduction in risk of environmental liability is factored in.
Improve access to capital and lower capital and other business service costs	Lenders tend to favor companies with an EMS because an EMS indicates better overall management and likelihood of business sustainability: They tend to lend to those companies first and at better rates. Environmental due diligence is now routine in mergers and acquisitions, and studies have shown that companies with EMSs tend to fare better in the stock market than competitors without EMSs. Many insurance companies also now prefer commercial clients with EMSs and, depending on the nature of the insurance, may offer lower premiums than they offer similar businesses without EMSs.
Improve market access	Governments, consumers, and major business customers increasingly demand that their suppliers demonstrate a commitment to environmental responsibility. As a result, companies with EMSs have access to some markets and specific customers denied competitors without EMSs.
Achieve better public, community, and government relations	A well-run EMS encourages dialogue and responsiveness to the concerns of outside parties. Some EMSs, such as the European Union's EMAS, require an active stakeholder outreach program. But even when that is not the case (ISO 14001, for example, does not require an outreach program), just having a productive EMS in place commends a company to its stakeholders as one that is environmentally responsible.

A sound EMS, whether based on BS 7750, EMAS, or ISO 14001 alone, will be productive. It will generate impressive improvements in environmental performance, operational efficiency, and environmental risk reduction for a company. Any productive EMS will provide a company with all the benefits listed in Table 1–3. But when a company's EMS has been certified to be in conformance with the requirements of the ISO 14001 international EMS standard, it tends to reap additional benefits. These additional benefits are especially in the areas of improved capital access, improved market access, and improved public image. Additional benefits accrue in these areas because "ISO 14001 certification" is acquired as the result of an independent audit by an official outside EMS auditor and therefore taken as proof that the company has a sound EMS in place. In other words, ISO 14001 certification gives the company's EMS visible legitimacy.

3.5 A Review and What Comes Next

An EMS is a system for continually improving the environmental performance, and thereby the business performance and sustainability, of a company or other organization. It is a tool that enables a company to address its environmental performance based on the way it operates, rather than just checking the status of its compliance with a collection of unrelated and unranked environmental standards. It is a means for top management to proactively, strategically, and comprehensively deal with the way the company interacts with the natural environment, in accordance with basic principles of good management. It is a tool that enables an enterprise to make good environmental performance a source, rather than a consumer, of revenues; and it elevates environmental management to the same plane as all other business components that are fundamental to strategic planning, growth, and sustainability for a corporation.

In concept, an EMS is a simple tool, comprising the same steps as a classical planning and implementation model made into a closed loop of continual improvement. The ISO 14001 international EMS standard is built around a planning and implementation model. The parts of the model correspond to (a) establishing the company's environmental policy; (b) identifying priority environmental problems, setting environmental targets for improvement with respect to those problems, and planning and carrying out programs of action to achieve the targets; (c) establishing the EMS infrastructure to support implementation of those programs of action and environmentally responsible behavior of the company in general; (d) measuring, monitoring, evaluating, and revising to improve the performance of the EMS and the environmental and economic performance of the company as a whole; and (e) reviewing, evaluating, and decision making by top management toward continual improvement of the EMS and the environmental and economic performance of the company.

The EMS generates environmental management programs that, if well thought-out and well implemented, result in improved environmental performance and regulatory compliance on the part of the company. If environmental management programs have been well thought-out and well carried-out, they also result in reduced costs of production, reduced regulatory penalties, and reduced risk of environmental liability, far into the future.

The aim in this chapter has been to provide a primer on EMS and ISO 14001. A rich literature elaborates the philosophy, full implications, and precise requirements of each element of ISO 14001 and provides approaches and specific tools and techniques to help with implementing an EMS. Readers who want to know more about EMSs in general and ISO 14001 in particular can refer to *Green Profits: The Manager's Handbook for ISO 14001 and Pollution Prevention* (London: Butterworth–Heinemann, 2001) and *Achieving Environmental Excellence: Integrating P2 and EMS to Increase Profits* (Washington, DC: Government Institutes, 2003). YES, an enormous volume of helpful material on EMS and ISO 14001 exists on the Internet.

The balance of this volume provides practical tools and knowledge to assist in the implementation of an EMS that complies with ISO 14001. Discussions and recommended approaches provide a method of self-assessing the status of an environmental management system; explains the steps in implementing an EMS, the purpose, techniques, and tools for each step; and offers guidance in EMS terminology. In Chapter 5, the reader will find a discussion on the methodology applied to conducting an initial environmental review, along with details on how to apply the software accompanying the publication to ranking the environmental aspects of the facility.

4

Components of an Environmental Management System

Introduction

There are seven components to an EMS:

1. Commitment and environmental policy.
2. Initial environmental review.
3. Planning the environmental policy.
4. Implementing the environmental policy.
5. Measure and evaluation.
6. Audits and reviewing.
7. External environmental communication.

Figure 4–1 summarizes these components for the reader. This chapter provides a description of each component. The discussions are intended to acquaint the reader with their terminology, purposes, and objectives.

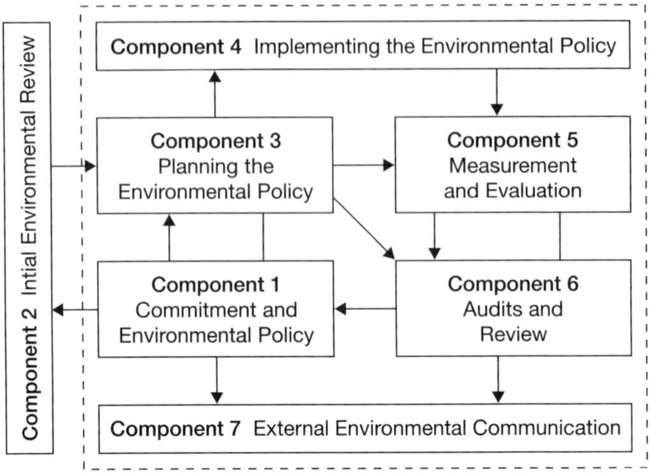

Figure 4–1 Components of an EMS and how they interact. (Modified from A. Sturm, "Implementing an Environmental Management System," Basel, Switzerland: Ellipson Management Consultants, 1998.)

4.1 Commitment and Environmental Policy

An EMS must have the commitment of top management to be effective. Without a formal commitment, the system will have neither credibility nor effectiveness. Commitment carries along with it time, money, staff, and energy—in short, the resources to support the policy.

The policy is a formal document with two parts. One part explains the rationale for the company to make the commitment. The second part states the overall intentions of the company regarding the management of the way the company interacts with the natural environment and the public. The policy contains a commitment to prevent pollution, comply with all applicable environmental regulations, and implement a process of continuous improvement. The following are excerpts from policy statements published by some well-known corporations.

Corporate Environmental Policy Statement of the Pepsi-Cola Co.

Pepsi-Cola has a strong history of caring for the environment. Our goal is to have the least possible impact on the environment, and so far we have been very successful. Among our environmentally friendly initiatives are the "Please Recycle" message on each and every container that we make, and our contributions to a number of programs designed to stop littering and promote recycling.

Pepsi-Cola is committed to being an environmentally responsible corporate citizen. In our offices, we use our resources wisely and actively recycle cans and paper. We not only strive to meet local and state environmental laws, but also try to exceed them by being industry innovators. For example, in 1992 Pepsi-Cola replaced its can holders with plastic ring connectors. Using a break-apart concept, these rings snap when cans are removed from the connectors, greatly reducing the risk of entanglement for wildlife. In addition, photo-degradable additives break down these connectors into small particles when they are exposed to sunlight, further reducing the likelihood of any negative environmental impact.[1]

Corporate Environmental Policy Statement of General Motors of Canada

As a responsible corporate citizen, General Motors of Canada (GMCL) is dedicated to protecting human health, natural resources, and the local and global environment, in accordance with the Environmental Principles of General Motors Corporation. This dedication reaches further than compliance with the law to encompass the integration of sound environmental practices into our business decisions. This policy is based on the integration of risk-based, cost-effective management practices into site activities with the aim of continually improving environmental performance.

GMCL is committed to assess the environmental impacts of its activities, products or services as a basis for its environmental management programs and to reduce wherever practicable these impacts through the establishment of appropriate objectives and targets. In particular, GMCL will strive to achieve the following objectives through continued execution of our Environmental Management System:

1. Comply with all applicable environmental laws and regulations, and other requirements.
2. Assign management responsibility for the environment in all areas of our facilities and ensure that all employees are aware of their individual responsibilities for acting in accordance with

[1]http://www.pepsi.com/help/faqs/faq.php?viewall=yes&category=community_info.

this policy, while providing effective information and training to encourage individuals to contribute effectively.

3. Practice effective prevention of pollution by seeking practicable ways to reduce energy and water consumption, and reducing discharges to air, water, and land.
4. Maintain communications with our local community, legislators, regulators, and other organizations with an interest in our environmental performance.

In accordance with our Environmental Management System (EMS) requirements, we will regularly review environmental performance and our facilities' objectives and targets to assess progress toward continual improvement. This policy statement will be periodically reviewed to ensure its continuing suitability, be available to the public, and communicated to all GMCL employees.[2]

Corporate Environmental Policy Statement of Exxon-Mobile Corp.

It is our policy to conduct our business in a manner that is compatible with the balanced environmental and economic needs of the communities in which we operate. We are committed to continuous efforts to improve environmental performance throughout our operations worldwide.

Our policy is to:

- comply with all applicable environmental laws and regulations and apply responsible standards where laws or regulations do not exist;
- encourage concern and respect for the environment, emphasize every employee's responsibility in environmental performance, and ensure appropriate operating practices and training;
- work with government and industry groups to foster timely development of effective environmental laws and regulations based on sound science and considering risks, costs and benefits, and effects on energy and product supply;
- manage our business with the goal of preventing incidents and of controlling emissions and wastes to below harmful levels; design, operate, and maintain facilities to this end;
- respond quickly and effectively to incidents resulting from our operations, cooperating with industry organizations and authorized government agencies;

[2]http://www.gmcanada.com/inm/gmcanada/english/about/Environment/environment.html.

- conduct and support research to improve understanding of the impact of our business on the environment, to improve methods of environmental protection, and to enhance our capability to make operations and products compatible with the environment;
- communicate with the public on environmental matters and share our experience with others to facilitate improvements in industry performance;
- undertake appropriate reviews and evaluations of our operations to measure progress and to ensure compliance with this policy.[3]

While there are similar commitments among these sample policy statements, each is distinct and uniquely associated with the respective company. Both the public and employees of each company can relate to a policy statement as a formal commitment that is specific to the company.

As with any official statement made by a responsible corporation, the policy statement must be communicated to all members throughout the company and should be accessible to various stakeholders, such as the public, investors, regulators, and other interested parties. Publicly traded companies, for example, make it a point to include their corporate environmental policy statement in their annual financial reports.

An environmental policy is the basis for developing a consistent structure for environmental objectives and targets. In this manner, a company can focus on high-priority areas, such as energy reduction, solid waste reduction, the phasing out of CFCs, the elimination of asbestos, and making the transition to biodegradable packaging are but a few examples.

4.2 Initial Environmental Review

An initial environmental review (IER) is the first action taken before an organization can plan and implement its environmental policy. The review provides a detailed overview of the environmental issues a company deals with. An IER will result in a baseline

[3]http://www.exxonmobil.com/corporate/Citizenship/Corp_citizenship_enviro_policy.asp.

characterization of how the company manages its environmental issues, which can be used as a basis to identify areas of deficiency and noncompliance. Once deficiencies and noncompliance concerns are identified and organized on a priority basis, environmental management initiatives can be devised and implemented to systematically eliminate the gaps. I recommend that the IER be implemented in three phases.

Phase 1 is aimed at *document organization and data collection*. A carefully planned audit questionnaire should be used as a guide to obtaining and organizing information on the current environmental management practices. Based on IERs conducted with international teams, a generalized audit questionnaire has been developed. The audit questionnaire, along with a description of a software product that will aid in performing the IER, are provided in Chapter 5. The software is contained in a CD in your book's jacket. It is recommended that you read over Chapter 5 when using the CD for the first time. The chapter provides an easy to follow example.

Phase 2 is *analysis*, with a focus on capturing immediate opportunities. As I recommended in previous publications (*Green Profits* and *Achieving Environmental Excellence*), the IER should not limit its focus to compliance-related issues. The intent of the EMS itself is not simply to bring a company into compliance but to reduce pollution at the sources on a continual basis. Hence, both the data collection phase and, especially, the analysis phase should include a balanced focus on identifying immediate pollution prevention (P2) opportunities. These will be mostly, if not entirely, P2 housekeeping actions and require relatively small investments in instrumentation, process controls, readily implemented process changes, simple QA/QC steps, or possibly tighter inventory practices. Later on, the company will develop a carefully considered P2 investment strategy. But, capitalizing on immediate P2 opportunities can provide powerful education and motivation among employees for the company's program, and the cost savings can even help offset the costs of implementing the EMS program.

Phase 3 is to develop *action plans* based on a further analysis of the IER data. The data and information collected represent a baseline characterization of the practices currently relied on by the company. The analysis is extended under this phase to benchmark the practices against industry and regulatory standards. As an example, a refinery that has a history of odor complaints or violations from air discharges has a deficiency compared to industry recommended practices as well as in meeting its compliance obligations in a consistent manner. The degree to which the practice or aspect of the operation is deficient (or produces negative impacts on the natural environment and public) should be assessed and given a score based on severe, moderate, or low to no negative impacts. This enables the company to develop a priority list of the aspects that should be addressed first. Once a ranking of the negative environmental aspects has been developed, targets for improvements can be established and environmental management programs (also known as *environmental action plans* or *environmental management initiatives*) can be formulated and recommended to top management for implementation. Chapter 5 provides further explanation by way of an example.

4.3 Planning the Environmental Policy

4.3.1 General

Proper planning requires knowledge of the company's interactions with the environment and public. This includes an understanding of legal and other obligations required of the company. From these, environmental objectives and targets are defined and the initial environmental management programs recommended from the IER can be finalized to meet the targets.

Environmental management programs should be specific. They should define the targets and objectives of the company clearly, establish the means and resources by which the targets can be met, and be placed on a schedule.

The environmental management system elements concerning the planning stage based on ISO 14001 include the following:

- Identification of environmental aspects.

- Evaluation of the resultant environmental impacts.
- Consideration of the company's legal requirements.
- Formulation of the environmental policy.
- Establishing internal performance criteria based on well-defined metrics.
- Establishing environmental objectives and targets.
- Establishing and implementing the environmental management program(s).

The environmental management program is a systematic, detailed roadmap that defines how the company will achieve its organizational objectives and targets. As a very first environmental management program, if your company does not have an EMS, then the program should be to implement one.

4.3.2 What Environmental Aspects Are

The term *environmental aspect* refers to the specific interactions that the company's operations have with the surrounding environment. A simple way to identify environmental aspects is to consider how materials and energy flow through the manufacturing process. Figure 4–2 illustrates the classic black box model used for material balancing purposes. By identifying the individual streams or actions resulting in the waste stream outputs, the negative environmental aspects become apparent.

Examples of negative environmental aspects are shown by the photographs in Figures 4–3 through 4–11. Negative environmental impacts fall into many different categories. They may be chronic health risks for a residential community through long-term exposure to air emissions, damage to aquatic life from contaminated storm water runoff, impairment to the quality of groundwater from chemical spills, off-site property damage from unplanned and chronic low-level emissions, worker accidents or health impairment due to on-site exposure to chemicals, or direct monetary losses from violations that lead to citations, fines, and penalties. These, in turn, hit a company in its pocketbook from both direct costs and indirectly from the diversion of resources, including but not limited

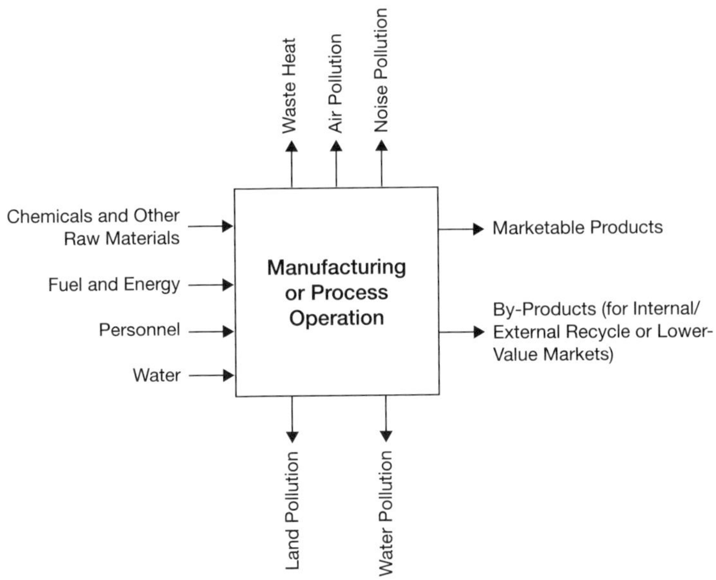

Figure 4–2 Black box model for any manufacturing facility.

to legal efforts to address them. They can lead to less tangible impacts, such as negative public image, which in turn can lead directly back to the corporate pocketbook by the boycotting of the company's products and services by a concerned public, loss of investor confidence and depreciation in stock values, and the raising of insurance premiums by insurance carriers because of a high-risk reputation. Negative impacts can add to present-day liabilities as well as future ones from contaminated soils and groundwater, which will have to be addressed by an environmental impact assessment, environmental sampling, cleanup, and possibly long-term monitoring. And, there may be negative impacts from lost opportunities because land areas require long-term remediation from contamination; hence, the development of land is restricted from industrial impacts. Negative impacts also may stem from aspects that relate to sustainability. In countries where water is limited, as in the Middle East, excessive use of water for industrial purposes along with poor management of wastewater have negative impacts on entire industry sectors.

Figure 4–3 Example of a negative environmental impact, oil spills from unloading crude oil from a tanker.

Negative environmental impacts can be more global in nature. Industrial activities have negatively influenced climate change. While entire governments, including the United States, act in an irresponsible manner with regard to this aspect by refusing to acknowledge and support the Kyoto Protocols, some companies are concerned and have formal management plans to address the negative impacts of those environmental aspects that contribute to global warming.

Figure 4–4 Example of a negative environmental impact, product losses and fugitive emissions from leaking flanges.

Figure 4–5 Example of a negative environmental impact, stockpiled asphalt and mixed wastes leading to long-term liabilities from soil and possible groundwater pollution as well as violation of Resource Conservation and Recovery Act rules.

Figure 4–6 Example of a negative environmental impact, contaminated surface runoff leading to off-site contamination and possible groundwater pollution.

Figure 4–7 Example of a negative environmental impact, poor housekeeping at an asphalt plant. Site contamination leads to soil and groundwater contamination, unsafe work conditions, fugitive emissions, and product losses.

Figure 4–8 Example of a negative environmental impact, a leaking pump resulting in fugitive emissions, product loss, and site contamination.

Most companies and their operating facilities have a large number of environmental aspects to manage. For large and multinational corporations, environmental management encompasses multiple sites. Establishing policy and environmental management plans requires planning and management of the aspects. To set objectives and define management plans, an analysis of the exposure a company has to different environmental aspects must be performed. A recommended approach to the identification of significant environmental aspects is accomplished by developing an environmental exposure portfolio, which results in a set of priorities. An analysis of the exposure portfolio points out the objectives of the environmental policy and the management plans. This is the focus of the third phase of an IER.

Figure 4–9 Example of a negative environmental impact, fugitive emissions and steam losses from leaking valves as a result of an ineffective preventive maintenance program.

Figure 4–10 Example of a negative environmental impact, long-term site contamination from poor practices in tanker loading operations.

Figure 4–11 Example of a negative environmental impact, a serious safety violation is the compressed gas cylinder in the middle of the access way.

4.3.3 Legal and Other Requirements

The legal and other regulatory requirements of a company should be assessed from the standpoint of both production-related environmental regulations and product- and service-related environmental regulations.

Production-related environmental regulations concern operating and manufacturing operations. Product- and service-related regulations concern marketing and R&D functions within the company. Both should be examined from three points of view:

1. In terms of the relevant environmental regulations. This helps define the *targets* that the company needs to achieve to be in compliance.
2. In terms of the current situation. In other words, what is the company's *baseline* in terms of compliance?
3. Finally, what are the *gaps*? Does the company comply with regulations and which ones does it not comply with?

Defining the *targets*, *baseline*, and *gaps* provides a basis for developing management plans.

The three questions can be addressed as a part of the IER through an audit questionnaire. An audit questionnaire segment is described in Chapter 5. A well-planned and -executed audit assists in organizing the company's information about current and also future regulations and liabilities. The information gathered includes

- Review of environmental regulations:

  The name or title of the regulation.

  The date of the regulation.

  The date of pending regulations.
- Compliance status within the company:

  Noncompliance.

  Expected to comply.

  100% compliance.
- Applicability to production and products.
- Responsibility: person's name and function.

By examining information on both current and foreseeable future regulations and both tangible and not so tangible liabilities, the company can develop a proactive management approach. A proactive approach enables the organization to develop and assess alternative ways to ensure uninterrupted compliance and meet future compliance challenges. This will result in continued savings to the company.

4.3.4 Objectives and Targets

The company's environmental policy is based on a commitment to continual improvement. Implementing a system that addresses objectives and targets ensures that the organization controls and guides the continuous improvement process. Objectives must be specific and the targets themselves measurable to be effective and in conformance with ISO 14001. The most-effective way to establishing objectives is by performing a careful analysis of the company's exposure to the environmental aspects identified in the IER.

4.3.5 Devising an Environmental Management Program

An environmental management program is the roadmap the organization will follow toward achieving its environmental objectives and targets. It is a document that provides the details of what must be done, by whom, how, and when for each of the defined objectives and targets. The objectives and targets themselves must be ranked in priority at the start. When devising the environmental program,

- Management must designate responsibility for achieving the objectives and targets at each function and level within the organization.
- Management must provide the means for fulfilling the objectives and targets. In a practical sense, this means allocating resources such as people, skills, technologies, financial resources, and the like. If the company is going to truly commit to implementing an EMS, then in short, management needs to put the corporation's money where its mouth is.
- A time frame in which objectives and targets will be achieved needs to be established. The implementers must be challenged to meet a realistic implementation schedule with the resources that are made available.

The environmental management program should be integrated into the company's overall strategic plan. After all, the idea behind an EMS is that it elevates environmental management to the same level of strategic business planning. In fact, no company can afford not to integrate environmental strategies into the overall business strategies and key operations aspects that are traditional in the corporate world of thinking.

It is standard practice that a site will have one environmental program. However, there are no rules against establishing additional programs as needed. New developments, partnering relationships, or new or modified products, processes, or services could justify the need for additional programs. The word to bear in mind is *flexibility*. Since the basic approach to an EMS is continual improvement by the plan-do-check-revise/act model, changing conditions not only may mandate changes to the basic environmental management program but call for additional programs to enhance the core activities.

The program should be a physical document. In fact, EMAS requires that the program be in written form and authorized at the highest level within the corporation. So that everyone in the organization is on the same page, the program should be distributed to everyone within the organization. It is also sound practice to distribute the program to contractors and subcontractors working at the site. The scheme devised in the environmental management program essentially requires that all engaged on the site in question conform to the site's policy.

Within the spirit of the EMS, environmental management programs should be periodically reviewed and regularly revised to reflect changes in the company's objectives and targets. Along these lines, it is obvious that performance monitoring using properly defined metrics be applied so that the progress of the program can be demonstrated as well as revised, if necessary. To accomplish this, a detailed description of any measures planned for the future or any measures likely to be taken in any particular circumstances is needed.

4.4 Implementing the Environmental Policy

An organization must develop the necessary capabilities and support mechanisms to achieve its environmental policy, objectives, and targets. Only then can an effective implementation be achieved. This often means that the policy will be implemented in stages. An effective stagewise implementation program depends on a number of factors: to name a few, the level of awareness of the environmental requirements of the organization, identification and proper priority ranking of the critical environmental aspects, management's expectations and the perceived benefits from the policy and EMS, and of course, the resources that are available. Key elements of the implementation stage are

- Structure and responsibility.
- Training, awareness, skills, and competence levels within the organization.
- Internal communication.
- EMS documentation.
- Operational control aspects.
- Emergency preparedness and response.

For *structure and responsibility*, the roles, responsibilities, and levels of authority designated to personnel whose activities have an impact on the environment require clear definitions. Empowerment of employees is a key factor in the ability to be successful and decisive during the implementation phase. This includes the effectiveness of key personnel in both documenting and communicating policy to all members of the organization. The right type and level of resources for the implementation and maintenance of the EMS must also be well structured. Most organizations appoint a special management representative, who has the responsibility and authority to enforce the EMS requirements. This person's responsibility is to report on the environmental performance of the organization directly to senior management.

For *training, awareness, and competence*, the entire organization must be made aware of the environmental policy, the environmental management programs, and the impacts of individual departments and individuals on the environment. Employees need to be trained by competent, qualified trainers and with meaningful programs, and in some instances, competency testing should be required. The following example may help to illustrate this point.

The Ellsworth Superfund site in Downers Grove, Illinois, is an industrial park where a number of large companies perform degreasing operations to clean metal parts. The businesses are diverse, but all the companies involved in the suit are large, experienced organizations. The groundwater has been badly contaminated with chlorinated solvents including trichloroethylene (TCE), 1,1,1-trichloroethane (TCA), tetrachloroethene, methylene chloride, and several other chlorinated chemicals. Despite each of the defendants in this matter being a large corporation that conducted right-to-know training under the Hazard Communication Act, key personnel did not understand how to interpret information on material safety data sheets (MSDS), did not have an understanding of the toxic and chemically aggressive nature of chlorinated solvents, and were not trained to a satisfactory level of competency in spill prevention and control. While a great deal of the environmental damages at this site may be attributed to historical practices, poor environmental management practices, comparable to 25 years ago, persist among the facilities. This is an example where companies have met their compliance obligations by providing training as required under the Hazard Communication Act, and yet the level of training and competency of the employees was inadequate. Companies have an obligation to exceed their minimum legal requirements, not only to protect the public and environment but also to protect their shareholders and investors. Companies that meet their minimum legal requirements may still be categorized as acting irresponsibly, because smart businesses recognize that merely meeting a legal statute does not ensure that environmental damages are not occurring. Companies that focus only on meeting their minimum compliance obligations may still face costly and long-range liabilities from environmental damages.

Employee training must go well beyond awareness. Training must be in-depth and targeted to those employees whose jobs have the greatest impact on the company's activities with respect to the environment. The training not only needs to be tailored to the most appropriate subject areas but provided at the level most appropriate to the educational and experience backgrounds of the employees. It is inappropriate to have someone who is experienced in safety management and expect that individual to assume environmental responsibilities without having first invested in proper training, education, and on-the-job work assignments that prepare him or her to assume such a role. Yet, time and time again, I observe this mistake among corporations.

For *internal communication* between various levels and functions of an organization, procedures should be established and maintained. Communication tools vary and their success in part depends on the company culture. Examples of tools are bulletin board postings, internal newsletters, meetings and electronic mail messaging, maintaining a virtual knowledge center, and maintaining an environmental and health and safety scoreboard on the company's server. Communication needs to be a two-way process. The information should be understandable and adequately explained in a timely fashion. Communication is also a crucial tool by which management can demonstrate its commitment to the EMS and policy; to raise employee awareness; to create an open dialogue with employees to address concerns and questions about the organization's activities, products, or services; and to just plain keep everyone informed. Communication tools enable the results of EMS monitoring, audits, and management reviews to be communicated throughout the organization and especially to those who are responsible for performance.

While P2 may be the heart of an EMS, *EMS documentation* is the main artery. If your company seeks ISO 14001 registration, then the EMS must be well documented to ensure that external auditors can certify the management system according to one of the current EMS standards. And, even if certification is not sought, but rather the decision is to adopt self-declaration of ISO 14001, the EMS needs to be well documented to no less of a degree. At a minimum, the documentation must include a description of the basic elements of the system and their interaction. It must point to other documents that include process information; organizational charts; internal standards and operational procedures (SOPs, standard operating procedures); emergency preparedness plans; spill prevention, contingency, and countermeasure (SPCC) plans; and others. The documentation may be integrated and shared with the documentation of other systems implemented by the organization, such as the ISO 9000 quality documentation. It is highly recommended that effort be invested into creating and maintaining an EMS handbook.

EMS documents need to be reviewed, revised, and approved on a regular basis, so that current information is available on tasks and work activities being performed. To ensure against unintended use, a document retention policy should be implemented. The policy should be formalized and maintained in written form, just like any other policy document. This ensures that obsolete documents are removed from all points of issue and points of use. Document retention policies require careful planning, and the company's legal department should be consulted. Normally, documents and records (including electronic) should be retained in accordance with the statute of limitations. As examples, patent invention records are enforceable up to 50 years and purchase records have a statute of limitation of 7 years. While environmental regulations like the RCRA (Resource Conservation and Recovery Act) mandate that records be maintained for 5 years, the fact is that environmental liabilities can surface from activities that happened 20 or more years ago. Most companies I worked with as a consultant typically retain environmental records like waste manifests, air and water discharge permits, and even training records for 14 years or more. Even when a document retention policy is implemented and obsolete documents are destroyed, it is good practice to retain one copy of each document as a permanent record. This applies to SOPs, a summary of chemical purchase records, sales records, and perhaps others that the company's legal counsel can advise on. Consideration should also be given to document preservation and retrieval protocols. If resources are available, historical records can be scanned and archived electronically. This saves space and preserves the records.

Operational control aspects focus on making sure that the environmental policy is implemented and the objectives and targets achieved. To accomplish these, the organization must identify processes and activities that have a significant impact on the environment and ensure that these activities and processes are conducted as intended. Procedures to control operations such as documented work instructions to ensure conformance to the EMS; monitoring and control of important process characteristics such as effluent waste streams, air emissions, and waste disposal; site security practices and procedures; and many others are key operational aspects that need to be drawn into the framework of the EMS and become a part of the overall policy.

Additionally, the organization needs to establish procedures for verification of compliance to specified requirements and to establish and maintain records of the progress. The responsibility and authority for measures to deal with noncompliance and subsequent corrective actions also should be specified. It is necessary to identify those environmental aspects related to the goods and services the company relies upon and communicate the organization's requirements to suppliers and contractors.

Emergency preparedness and response may be argued as a part of the operational control aspects, but given the complexity and size of most refinery operations, it merits separate and distinct attention by the organization. It is crucial that effort be placed in identifying the potential for accidents and emergency situations and developing appropriate procedures to respond to these. Hazard and operability studies (HAZOPS) should be applied.

HAZOPS are hazard and operability studies, the technique of which has been used and developed over approximately four decades for "identifying potential hazards and operability problems" caused by "deviations from the design intent" of both new and existing process plants. Because of the high profile of production plant accidents, emphasis too often is placed on the identification of hazards to the neglect of potential operability problems. Yet, it is in the latter area that benefits of a HAZOP study are usually the greatest.

Safety and reliability in the design of any manufacturing plant, whether it be a chemical plant, a fertilizer manufacturing facility, or a refinery initially relies on the application of various codes of practice or design codes and standards. These represent the accumulation of knowledge and experience of both individual experts and the industry as a whole. Such application is usually backed by the experience of the engineers involved, who might well have been previously concerned with the design, commissioning, or operation of a similar plant. Although codes of practice are extremely valuable, it is important to supplement them with an imaginative anticipation of deviations that might occur, for example, because of equipment malfunction or operator error.

Essentially, the HAZOPS procedure involves taking a full description of a process and systematically questioning every part of it to establish how deviations from the design intent can arise. Once identified, an assessment is made as to whether such deviations and their consequences can have a negative effect on the safe and efficient operation of the plant. If considered necessary, action is then taken to remedy the situation.

This critical analysis is applied in a structured way by the HAZOP team. It relies on team members releasing their imagination in an effort to discover credible causes of deviations. In practice, many of the causes will be fairly obvious, such as pump failure causing a loss of circulation in cooling water to heat exchangers. However, the great advantage of the technique is that it encourages the team to consider other less obvious ways in which a deviation may occur, however unlikely they may seem at first consideration. In this way, the study becomes much more than a mechanistic checklist review. The result is a high probability of identifying potential failures and problems that had not been experienced previously in the type of plant being studied.

The HAZOP study process involves applying logic to uncovering potential problems. The results are recorded in columnar format under the headings Deviations, Cause, Consequences, Safeguards, and Action. While the terms listed in these columns may seem obvious to the reader, it is worthwhile stating their definitions.

- *Deviation.* The keyword combination being applied (e.g., Flow/No.).
- *Cause.* Potential causes that would result in the deviation occurring (e.g., Pump P1 blockage due to impurities in Dosing Tank TK5; this might be a cause of Flow/No.).
- *Consequence.* The consequences that would arise from both the effect of the deviation (e.g., loss of dosing results in incomplete malfunctioning of biological contactor in wastewater treatment unit BC1) and, if appropriate, the cause itself (e.g., cavitation in Pump P1, with possible damage if prolonged).

It is important be explicit in recording the consequences, in keeping with the spirit of the overall EMS. It should not be assumed that the reader at some later date will be fully aware of the significance of short, abbreviated statements. It is much better to add an explanation.

When assessing the consequences, one should not take any credit for protective systems or instruments already included in the design. For example, suppose the team had identified a cause of Flow/No. as being spurious closure of an actuated valve. It is noticed that there is valve position indication within the Central Control Room, with a software alarm on spurious closure. The team may be tempted to curtail consideration of the problem immediately, recording something to the effect of "Minimal consequences, alarm would allow operator to take immediate remedial action." However, further investigation might have revealed that the result of that spurious valve closure would be overpressure of an upstream system, leading to a loss of containment and risk of fire if the cause is not rectified within three minutes. Only then does it become apparent how inadequate is the protection afforded by this software alarm.

- *Safeguards.* Any existing protective devices that either prevent the cause or safeguard against the adverse consequences would be recorded in this column. For example, you may consider recording "Local pressure gauge in discharge from pump might indicate problem arising." Note that safeguards need not be restricted to hardware; where appropriate, credit can be taken for procedural aspects such as regular plant inspections. In this regard, an EMS is designed to ensure that plant inspections actually are carried out.
- *Action.* Where a credible cause results in a negative consequence, it must be decided whether some action should be taken. Consequences and associated safeguards are considered at this stage. If it is deemed that the protective measures are adequate, then no action need be taken; and words to that effect are recorded in the Action column. Actions fall into two groups:

  1. Actions that remove the cause.
  2. Actions that mitigate or eliminate the consequences.

  Whereas the former is to be preferred, it is not always possible, especially when dealing with equipment malfunction. However, always investigate removing the cause first, and only where necessary mitigate the consequences. For example, to return to the "Pump P1 blockage due to impurities etc." entry referred to previously, we might approach the problem in a number of ways:

1. Ensure that impurities cannot get into TK5 by installing a strainer in-line preceding the pump.
2. Consider carefully whether a strainer is required in the suction to the pump. Will particulate matter pass through the pump without causing any damage, and is it necessary to ensure that no such matter gets into BC1?
3. Install both a strainer and a differential pressure gauge across the strainer, with perhaps a high differential pressure alarm to give clear indication that a total blockage is imminent.
4. Fit a duplex strainer, with a regular schedule of changeover and cleaning of the standby unit.

The team that conducts the HAZOP study should consist of personnel with a good understanding of the process and plant to be reviewed. The group should ideally be separate and distinct from any EMS management team but should report through them. In a study in which both contractor and client are participating, it is desirable to maintain a balance between the two in terms of team membership so that neither side feels outnumbered.

Along slightly different lines, written emergency preparedness and response plans are not only obvious for meeting responsible care practices followed by the industry as a whole, they are a legal requirement. One U.S. federal legislation where they are required is under the Emergency Planning and Community Right-to-Know Act (EPCRA), Title III of the Superfund Amendments and Reauthorization Act of 1986. EPCRA is intended to improve local community access to information about chemical hazards and improve state and local emergency response capabilities. State right-to-know acts also require reporting various kinds of chemical hazard information. EPCRA has three main objectives: (a) to bolster local emergency planning efforts, (b) to improve emergency notification in the event of a release of hazardous chemicals, and (c) to develop a baseline on routine chemical releases into the environment. To meet these objectives, EPCRA created four types of reporting obligations for facilities that store or manage specified listed chemicals. In 1994, over 300 more chemicals were added to the list of chemicals for which reporting is required. All information submitted pursuant to EPCRA regulations is publicly accessible, unless protected by a trade secret claim.

- *Notification of extremely hazardous substances.* EPCRA §302 requires facilities to notify the state emergency response commission (SERC) and the local emergency planning committee (LEPC) of the presence of any "extremely hazardous substance" if it has the substance in excess of the specified "threshold planning quantity." The list of such substances is in 40 CFR Part 355, Appendices A and B. It also directs the facility to appoint an emergency response coordinator.
- *Notification during releases.* EPCRA §304 requires facilities to notify the SERC and the LEPC in the event of a release exceeding the "reportable quantity" of a CERCLA (Comprehensive Environmental Response, Compensation, and Liability Act) hazardous substance or an EPCRA extremely hazardous substance. EPCRA extremely hazardous substances and reportable quantities are listed in 40 CFR 355.
- *Emergency planning.* EPCRA §311 and §312 require facilities to notify SERC, LEPC, and the local fire department of all hazardous chemicals for which the Occupational Health and Safety Administration (OSHA) requires material safety data sheets. The facility must submit either the MSDSs or a list of the substances for which MSDSs are maintained. If a list is submitted, hazardous chemical inventory forms (also known as Tier I and II forms) must also be submitted. A Tier I form provides information about hazardous chemicals grouped by hazard category. A Tier II form provides information about each specific hazardous chemical. This information helps the local government respond in the event of a spill or release of the chemical. These requirements are found at 40 CFR 370, Hazardous

> Chemical Reporting: Community Right-to-Know. Toxic Release Inventory, EPCRA §313, requires manufacturing facilities included in SIC codes 20 through 39 that have 10 or more employees and manufacture, process, or use specified chemicals in amounts greater than threshold quantities, to submit an annual toxic chemical release report to the Environmental Protection Agency (EPA). This report, commonly known as Form R, covers releases and transfers of toxic chemicals to various facilities and environmental media and allows the EPA to compile the national Toxic Release Inventory (TRI) database. These requirements can be found at 40 CFR 372, Toxic Chemical Release Reporting: Community Right-to-Know.

The Hazard Communication Standard (HCS) is another legislation that mandates emergency preparedness and response. The Occupational Safety and Health Act of 1970 established the Occupational Safety and Health Administration (OSHA) within the U.S. Department of Labor. The original act included language to the effect that employees should be apprised of all hazards to which they are exposed on the job. In the early 1980s, OSHA implemented this instruction by enacting the Hazard Communication Standard as 29 CFR 1910.1200. The HCS became effective in 1986. A fundamental premise of the HCS is that employees who may be exposed to hazardous chemicals in the workplace have a right to know about the hazards and how to protect themselves. For this reason, the HCS is sometimes referred to as the Worker Right-to-Know legislation or more often just as the Right-to-Know law. Although the original HCS applied only to the manufacturing industry, subsequent court challenges have modified the scope of the law so that today the HCS applies to nearly all sectors of the workforce. The Hazard Communication Standard sets forth guidelines and requirements in six areas:

1. *Chemical labeling* (29 CFR 1910.1200(f)). This guideline requires that all chemicals in the workplace be labeled. The information that must be present includes the name of the chemical and warnings about any hazards the material may present. This requirement may be implemented in a variety of

ways. Two such systems are the National Fire Protection Association (NFPA) Hazard Identification System and the Hazardous Materials Identification Guide.

2. *Material safety data sheets* (29 CFR 1910.1200(g)). An MSDS is a document that gives detailed information about a material, including any hazards associated with the material. MSDSs must be immediately available to employees at locations where hazardous materials are used.
3. *Hazard determination* (29 CFR 1910.1200(d)). This guideline says that the employer must identify and maintain a list of all hazardous chemicals used in the workplace.
4. *Written implementation program* (29 CFR 1910.1200(e)). This guideline mandates that the employer develop a written plan, the hazard communication program, detailing how the requirements of the HCS are implemented by the employer.
5. *Employee training* (29 CFR 1910.1200(h)). This guideline requires that the employer provide the employees training covering the handling of hazardous materials, the use and interpretation of both MSDSs and HAZCOM labels, and information about the HCS.
6. *Trade secrets* (29 CFR 1910.1200(i)). This guideline sets forth the conditions under which a manufacturer may withhold information about a material, and the conditions under which such information must be divulged to health care providers.

4.5 Measuring and Evaluating

The value of an EMS to a company is its ability to tailor the flow, control, and management of information on the environmental aspects that the facility is expected to manage responsibly. Once the environmental policy has been established and implemented, the organization, and in particular senior management, needs to be able to monitor and evaluate the environmental measures implemented under the environmental management plans and assess their impact on the environment. This is known as *environmental performance evaluation.*

An organization applies an environmental performance evaluation to assess the organization's performance against environmental targets and objectives and against applicable regulations. Responsibility and authority need to be defined to deal with noncompliance issues within the EMS. This includes specifying actions to be taken to correct undesirable situations and prevent future noncompliance and exposure to liabilities.

A range of terms are used to describe how best to measure and evaluate, but the terminology all have the same meaning. Rather than confuse the reader, I use consistent terminology that is most often applied by EMAS implementers. From a practical standpoint, implementing an EMS under EMAS is the same as implementation under the U.S. approach to ISO 14001.

The first step involves the development of an *environmental effects register* (EER). An EER is an inventory of environmental-related documents. By organizing information, two objectives are met. First, the company establishes a basis for managing the environmental aspects and its business as a whole in an environmentally responsible way and provides a basis for monitoring and assessing performance through information control. Second, the information gathered, if done systematically and diligently, can be organized in such a manner that it can be verified by an external auditor. This enables the company to seek formal registration of its EMS. The general categories of information needed by management for control and monitoring purposes range from environmental data, to regulations, to environmentally-induced financial information.

4.5.1 Developing an Environmental Effects Register

Recording and maintaining environmental data, regulations, and even environmentally-induced financial information is nothing new. This always has been an activity within companies, but historically such information has not been applied as a basis for effective decision making. Companies that operate without an EMS heavily discount such information in their strategic planning or simply do not consider it all. Often information on environmental cost accounting is buried in other financial information, but for most companies, it does exist. Historically, most companies simply never bothered to track or monitor the information and apply it to facility investments and long-range planning.

In this regard, an environmental manager who functions under an EMS finds that collection and preparation of physical environmental information across a number of categories is a major part of the job description. This is a significant departure from traditional environmental management practices, which focus more on monitoring emissions, meeting reporting requirements under permits to discharge, and responding to citations and violations when they occur.

When building the EER it is important to decide from the start the time frame over which critical information should be recorded and monitored. Most companies elect to use the same time period as in financial accounting, one year. This has two advantages. First, it enables environmental data to be measured in unit of input per year or unit of output per year (analogous to an emission factor). This normalizes any comparisons to a base year, which enables performance evaluations over time (e.g., in terms of unit of production or unit of crude processed compared against a base year). Overall environmental performance and even certain aspects of environmental performance, such as different pollution media, can be readily monitored in such a manner, including in terms of percentages of reductions in emissions. Second, it enables environmental data and information to be directly compared to financial indicators such as profits, losses, and costs and in terms of efficiency indices, such as cost of compliance per unit of production as compared on a year-by-year basis.

A second important decision that is needed in devising the framework for the EER is what organizational units should be included for the purposes of monitoring. By this, I mean that parts of operating units at a specific location can be defined as part of the organization while others can be excluded. An example of this is when a facility relies on an activity like cogeneration or perhaps enters into an off-site activity in which a waste is converted to a form of energy or recyclable venture. Another example is when a company operates more than one facility. The question reduces to whether or not organizational units that are monitored are defined on the basis of a legal corporate entity as opposed to specific locations, even within the same property boundaries of a single site. While some will argue that this is manipulation of environmental performance information by excluding a unit from the system, thus allowing better environmental performance reporting, nothing is wrong with the approach, provided the definition of the framework is transparent and stated from the onset. In fact, ISO 14001 allows for such flexibility.

Once the operational units are defined, the categories and specific types of data to be collected must be defined. Since no two refineries are the same, the effort requires some thought at the local management level. As a starting place, the refinery material output matrices in Chapter 2 can be used as a basis to develop an inventory of the environmentally relevant data that may be collected. It is important to bear in mind that the actual process of data collection will be spread over years, becoming more in-depth with each successive year. Hence, over time, the inventory builds. This approach enables an organization to fine-tune both its monitoring and management of the facility's environmental aspects using a scheduled and stagewise model.

4.5.2 Monitoring and Measuring Hardware

All equipment used for monitoring and measuring purposes must be accurate and calibrated on a regular basis. Not only does the facility need to be aware of what equipment, hardware, and instrumentation exist for pollution-monitoring purposes, but its age, its accuracy, the calibration procedures and frequency of calibration, the required maintenance and serviceability, SOPs, and so forth. All these are critical to managing environmental aspects. In short, a whole subuniverse of documents is important to maintaining and using the hardware and equipment important to monitoring and measuring pollution. Where this documentation exists, the age of the documentation, who has the responsibility for maintaining the various documents, and how these documents can be accessed are practical questions that need to be addressed as a part of the EMS.

4.5.3 Compliance Status

For management to monitor the compliance status of the company, information about the regulations and other requirements is needed. This can be addressed by first developing an environmental regulations register (ERR). The environmental manager needs to list all the pertinent regulations that apply to the facility, obtain copies of those regulations, and build a register that is accessible to the organization. Placing an ERR on the company server for all employees to access is a good idea. The ERR's role is to identify and serve as a depository of legal and regulatory requirements. From a practical standpoint, information is needed at two levels: production-related environmental regulations and product- and service-related environmental regulations. The former is concerned with the production departments, while the latter is concerned with the marketing and R&D departments.

Two questions that should be addressed:

1. With which environmental regulations should the facility be concerned? The environmental manager and management should view these as the "target."
2. Is the current status of the organization known? Surprisingly, many companies struggle to answer this question, which is justification enough for an EMS. The organization's status or "actual" situation needs to be compared against the target to define the appropriate actions for meeting compliance and improving undesirable situations.

The environmental manager can start off building the ERR very simply on an Excel spreadsheet following the format listed in Table 4–1.

All states maintain an EPA or department of environmental protection website. These are the best sources to obtain information from.

To assess the organization's compliance status, an analysis of the internal implementation of all the applicable regulations and its activities, products, and services is needed. For each regulation listed in the ERR the following two questions should be addressed:

Table 4–1 Recommended Format for an ERR

Register of Environmental Regulations			*Requirements*		*Applicable To*		*Responsibility*
Regulation	Date	Expected	Description	Relevant to Organization	Production Process(es)	Product(s)	Name and Function
. . .	. . .	. . .	. . .	. . .	. . .	. . .	. . .
. . .	. . .	. . .	. . .	. . .	. . .	. . .	. . .

Table 4–2 Recommended Spreadsheet Format for Assessing Legal Compliance

Regulation	*Dated*	*Expected*	*Compliance Status: Noncompliance/ Expected to Comply/ 100% Compliance*	*Applicable to: Production/ Product(s)*	*Responsibility*
. . .	. . .	. . .	. . .	. . .	. . .
. . .	. . .	. . .	. . .	. . .	. . .
. . .	. . .	. . .	. . .	. . .	. . .

1. Does the company comply with the regulation?
2. How well does the company comply with each of the requirements of the regulation?

Table 4–2 is a suggested format to be used in making the assessment. This can be set up simply on an Excel spreadsheet for tracking and analysis purposes.

In performing the analysis the company's regulatory status can be clearly established along with action plans to meet obligations by assigning the applicable regulations to one of the following groups:

1. Noncompliance with major environmental regulations.
2. Noncompliance with minor environmental regulations or certain statutes within regulations.
3. Current noncompliance but expected to comply in the near future.
4. 100% compliance with current environmental regulations.
5. 100% compliance with current and expected environmental regulations.

Groups 1 through 3 do not meet the overall requirement of an EMS, namely, to meet environmental regulatory requirements. The regulations listed in these groups require that immediate corrective actions be taken. Group 4 is equivalent to meeting the minimum requirement of the EMS. Group 5 meets the spirit of an EMS.

Using this approach, the company knows its current status, can define actions to bring the organization within compliance, and can identify both opportunities and threats arising from changes in regulatory requirements.

4.5.4 Environmental Cost-Accounting Data

Environmental cost accounting represents a body of information that an organization should be tracking closely. This information provides management with business metrics on the cost of compliance. At some point, the EMS's pollution programs will have significant positive impacts on environmental costs, driving them lower.

But, at the very early stages of an EMS implementation, management needs to go through a learning curve that can be accelerated by investing a little time in examining historical environmentally-induced cost data and dissecting it. For example, when accounting for waste disposal costs, the standard accounting practice is to simply record waste treatment costs, such as payments to the waste management contractor (i.e., the costs of disposal including transport). However, when one considers the internal costs of handling, management time, drums for storage, personnel effort in filling out manifests and filing reports, costs associated with space requirements for on-site storage, personnel for inventorying and inspections when wastes are stored for lengthy periods of time on-site, waste sampling, testing and profiling to meet reporting and transporter requirements, and insurance premiums, then the true costs for compliance are both significantly greater than the comptroller's expenditure reports reflect and quite enlightening to management.

A difficulty with tracking environmental-related financial information is that most organizations simply do not differentiate between capital expenditures and ongoing costs for environmental-related programs and other operating costs and site improvements, unless specific accounting instructions and protocols are established. By the same token, the savings accrued from P2 investments may be lost among other financial information unless monitored under separate accounts.

At the start, it is a good practice to instruct accounting departments to differentiate according to environmentally-induced financial information. This enables the organization to first develop baseline information on compliance and other environmental cost data. The environmental manager can devise the appropriate criteria whereby all costs and revenues related to environmental performance can be identified. Some obvious categories are waste disposal costs; spill cleanup costs; emission reduction costs; costs for pollution control equipment, air sampling, and water monitoring instruments; cost of raw chemicals for wastewater treatment; OSHA and related training costs; and site remediation costs. The list can be added to by including past fines and penalties from violations, cost for lost production time due to emergency shutdowns resulting from unplanned releases, cost of environmental consultants, cost of EIA (environmental impact assessments), and O&M (operation and maintenance) costs for pollution controls and monitoring instruments; further, estimated costs can be developed for legal costs resulting from fines, penalties, third-party legal suits, and personnel expenditures responding to public complaints. This list can be extensive, and it may not be possible to gather all pertinent information; for example, the cost of energy for pollution control and waste management or lost opportunity costs for space currently allocated for waste storage may be difficult to determine. These nonetheless are ongoing costs related to compliance, and their cumulative effects can represent significant outgoing cash flows that reduce profitability. Table 4–2 provides a list of environmentally-induced information categories that help the environmental manager and top management start developing a cost-monitoring program aimed at differentiating compliance-cost information.

Companies traditionally focus on compliance alone, and by doing so, they rely heavily on end-of-pipe treatment technologies. The consequence of this is that they ignore or do not recognize the true or total costs associated with pollution control and post-waste handling. To account for all of the costs we need to recognize that they fall into four distinct categories or tiers. R. T. McHugh (see *The Economics of Waste Minimization*, New York: McGraw-Hill Book Publishers, 1990) devised the following groups of costs:

Tier 1. Usual or normal costs.

Tier 2. Hidden or indirect costs.

Tier 3. Future liability costs.

Tier 4. Less-tangible costs.

Examples of each category follow.

Tier 1. The Usual or Normal Costs

Usual or normal costs include such costs to the operation as

- Pollution fees (i.e., fees paid for discharging pollutants within legally allowable emissions standards).

- Direct labor costs.
- Raw materials (e.g., feedstock chemicals, water).
- Energy.
- Capital equipment items.
- Site preparation for pollution control equipment.
- Equipment tie-ins and process modifications to accommodate pollution controls.
- Employee training.
- Permits to construct.

Tier 2. The Hidden or Indirect Costs

Hidden or indirect costs include such costs to the operation as

- Expenses associated with monitoring.
- Permitting fees (e.g., permits to operate, permit renewals, others).
- Environmental transformation costs (e.g., those costs associated with transforming a water pollution problem into a solid waste disposal issue. One example is a dewatering operation, such as a filter press to treat suspended solids in a wastewater stream. This strategy generates a sludge that must be disposed of. Another example is a wet scrubber used to capture particulates to address an air pollution problem. In this example, although the air emission problem is eliminated or reduced, a water pollution problem is generated with the use of this technology. The water pollution problem may also have to be addressed by separate treatment technologies, resulting in a sludge disposal problem).
- Environmental impact statements (depending on the nature of the project, EIAs can cost many thousands of dollars and require long lead times).
- Health and safety assessments.
- Service agreements.
- Legal costs.
- Control instrumentation.

- Maintenance and replacement costs.
- Reporting and record keeping.

Tier 3. The Future Liabilities

Future liabilities include such costs to the operation as

- Remedial action costs (site cleanup plus cleanup costs associated with third-party damages).
- Personal injury to workers.
- Health risks and injury to the public.
- More stringent compliance (environmental compliance is oftentimes a moving target, hence, a control technology used today may become obsolete a few years down the road).
- Inflation (e.g., this could affect higher feedstock costs and energy as well as waste disposal services).

Tier 4. The Less-Tangible Costs

Less-tangible costs include such costs to the operation as

- Consumer response and loss of confidence due to perceived poor environmental management (it is well recognized in technologically advanced societies that consumers and investors favor those companies that are environmentally conscious of their actions and have proactive environmental management programs).
- Employee relations (poor management of environmental issues places workers at health risks).
- Establishing and extending lines of financial credit (lending institutions will not extend loans to or provide favorable terms for credit lines if a company runs a high-risk operation from an environmental management standpoint).
- Adverse impacts on sales of property, mergers, and acquisitions (companies that are looking to partner or merge with others are not willing to take on the other firm's environmental liabilities; poor environmental management can bring enormous liabilities to investors and partner organizations).

- Higher insurance premiums (those companies that have a poor environmental track record represent a risk, and therefore, the potential loss of assets from fire or explosion or health risks to workers and the public are high. Insurers will not take these risks, or they will impose high premiums with very limited coverage).
- Becoming the focus of frequent environmental inspections and corrective actions (if environmental regulators recognize a recurring theme of noncompliance or accidents, then the facility will be singled out for numerous violations, fines, inspections, and corrective actions, all of which can result in significant costs to an operation, including interruptions in production schedules and legal costs).
- Less leverage in negotiating environmental fines and penalties (the environmental laws in technologically advanced nations are so complex, and operations are oftentimes fragmented, leading to innocent violations of environmental laws. If a company has had a history of poor environmental management, then it has little basis to negotiate for leniency on innocent violations. The problem is perceived as systematic).

Tier 1 and 2 cost components are usually identified in selecting pollution management strategies. If we look at these tier cost components alone, then it is easy to justify investments into control-based technologies. In contrast, Tier 3 and 4 costs are often not given a sufficient weighting (i.e., they are considered less important than more direct and tangible costs, usually because components are more difficult to assign values to), or they are simply not recognized. This is a mistake, because by attaching the proper importance and most-probable costs to these components, other strategies become apparent and more attractive.

Proper evaluation of Tier 3 and 4 cost components in assessing total costs and applying the results as a basis to selecting proper pollution management strategies requires the application of risk assessment principles. Since the ultimate goal is to arrive at cost-effective strategies that meet a compliance target, or to go beyond that target, risk assessment should be considered as part of life-cycle costing analysis (LCCA).

Understanding and dissecting the cost components within each tier are critical to making decisions on investing in the most-economical pollution management strategies. When companies focus their attention only on the obvious costs (Tiers 1 and 2), critical factors that could alter the decision on investments into pollution prevention may be missed. When their cost assessment includes such issues as worker health risks, cumulative sewer discharge fees, on-site wastewater treatment costs, and the cumulative costs associated with sludge stabilization, on-site staging of wastes, record keeping and waste reporting requirements, transport and disposal, then other alternatives that displace or eliminate the generation of wastes as a part of the process become much more attractive. When we consider also the potential future liabilities when the facility is sold, such as the costs that may be faced in remediating the property, then higher investments into strategies that eliminate pollution as a part of the operation are justifiable. This is the basis under which life-cycle costing techniques are applied to making logical investment decisions.

It should be obvious that collecting and recording data should start only after the system is defined and a basis has been established for making yearly comparisons. Data collection and recording the information is the last step in developing an EER. Once the system is established, the environmental manager or refinery manager can develop trend charts and begin identifying ways to capitalize on opportunities and reduce costs. This is where P2 opportunities can play an important role.

A final note is that, given the complexity of refinery operations and the large number of variables in process operations and pollution management, data collection and ultimately its analysis are major efforts. The plant and environmental managers need the proper tools to perform these tasks efficiently and in a timely fashion. In this regard, the expertise of IT (information technology) specialists should be consulted. There may or may not be appropriate off-the-shelf software suitable for managing large amounts of environmentally-induced cost information, which is only a portion of the overall information management requirements under an EMS. While a multitude of database management systems are commercially available, none seems to be acknowledged industrywide as appropriate in all cases. In my experiences as a consultant working with a number of international clients, I have observed more of a trend to rely on in-house software systems tailored to the needs of the organization. But, that in itself appears more of personal choice, driven partly by the perception among organizations that environmental information is sensitive and oftentimes should be kept proprietary for fear of it being misinterpreted by the public or laypersons. I personally believe environmental management should be transparent. It is in the best interest of both the company and the public at large. Regardless, an upfront investment is needed to identify, or develop in-house, the appropriate software tool, along with a learning curve for the staff to gain confidence and proficiency in applying it to the task at hand.

4.5.5 Combining Concepts with Pollution Prevention

The first overall objective of any pollution management strategy, and indeed an EMS, is to bring a facility into compliance with environmental regulations. Simply stated, each company faces only two choices:

1. Comply with environmental regulations and remain in business, or
2. Do not comply and face heavy fines, penalties, costly interruptions, or in the extreme, go out of business.

When a company approaches this objective with strategies based largely on end-of-pipe treatment technologies, it addresses only the legal requirements for staying in business and not growth. By repeatedly trying to meet more-stringent effluent emission standards through the use of more-advanced controls, a company is always on the defensive, that is, playing catch-up with compliance requirements. At the same time, this strategy at best meets only the minimum legal obligations. A good track record of compliance unto itself does not ensure that property assets, soil, groundwater, or off-site negative impacts from pollution are not occurring or have been over a long period of time. The reason this occurs is that the company's energy becomes focused on meeting the reporting and permitting requirements as opposed to reducing pollution impacts. In over 30 years of industry practice, I have encountered many facilities where pervasive contamination problems persist simply because normal chemical handling practices cumulatively over time contribute to groundwater contamination and off-site migration due to fate and transport mechanisms. These negative environmental impacts and the subsequent financial liabilities to companies can be avoided by eliminating pollution at the source and eliminating poor chemical handling and management practices or the use of outdated technologies.

P2 strategies achieve the same results as the application of control technologies (i.e., they help meet a compliance requirement), but P2 is more cost effective. The reason for this is that P2 strategies reduce Tier 1 and 2 costs through savings not achievable through control technologies, which are simply add-on process components that impose additional demands on energy, materials, and labor. The savings can be reinvested in modernization and improvements to operations or financing of the implementation of an EMS. This enables a company to grow its operations.

P2 strategies also reduce or eliminate costs associated with future liabilities (Tier 3) and less-tangible costs (Tier 4). In the ideal case, prevention eliminates waste and pollution altogether. Hence, issues like third-party damages or joint and several liabilities for off-site damages no longer are a concern.

When P2 strategies are employed, the owners and operators of facilities can exceed compliance requirements, because in the ideal case, zero waste and pollution is the ultimate goal. Another way of looking at this is that, when an organization focuses on continual reductions of waste and pollution, operating costs are steadily reduced over time and profitability increases.

P2 programs work best within the framework of an EMS. As I noted with my coauthor in *Green Profits* (Chapter 1): "An environmental management system is an approach . . . a tool . . . a set of procedures . . . a planned and organized way of doing things." In other words, an EMS is a systematic approach for a company to manage its environmental concerns. Most people tend to separate the functions of an EMS and pollution prevention in their minds or think of these as related only in the sense that P2 is the preferred way of addressing environmental aspects of a company's operations that have been identified through an EMS. But, in fact, a pollution prevention program must be fully integrated with an EMS to maximize the benefits a company can achieve through its EMS.

A P2 program consists of a systematic approach to identifying more cost-effective strategies and technologies for waste handling, then implementing those strategies and tracking their environmental and economic performances to establish new targets and goals, starting with the highest environmental priorities, then iteratively repeating the process for and applying lessons learned to successive pollution problems throughout the entire operation. Through the application of this iterative process, incremental savings are achieved with each new P2 activity. The savings are cumulative and, hence, over time, represent significant capital returns that can be reinvested into a company's operations. Successive rounds of continual P2 improvements are what constitute the basis for an EMS.

An overwhelming number of success stories illustrate the benefits of pollution prevention strategies. Many examples for a variety of industry categories are summarized in earlier publications devoted to this subject (N. P. Cheremisinoff, *Handbook of Pollution Prevention Practices*, London: Butterworth–Heinemann, 2001; and N. P. Cheremisinoff, and A. Bendavid-Val, *Green Profits: The Manager's Handbook for ISO 14001 and Pollution Prevention*, London: Butterworth–Heinemann, 2001). These case studies show distinct financial advantages to companies by identifying not only reductions in pollution and the costs associated with pollution management, but through reduced raw material consumption also energy savings, reductions in treatment and disposal of wastes, and reductions in labor associated with environmental management. Many P2 and waste minimization strategies, such as substituting toxic materials with safer alternatives, do not require process changes and, as such, are simple and cost very little to implement. The areas in which P2 has proven effective include the elimination and reduction of impacts from

- Treatment, disposal, and associated labor costs
- Wildlife and habitat damage
- Property devaluation
- Remediation costs
- Civil and criminal fines
- Permitting fees
- Insurance costs
- Process outages and disruptions

Case studies also testify to the fact that P2 benefits result in

- *Enhanced public image.* Consumers more-favorably view businesses that adopt and practice P2 strategies, and the marketing of these practices can assist in increasing a company's profits.
- *Increased productivity and efficiency.* P2 assessments have proven helpful in identifying opportunities that decrease raw materials use, eliminate unnecessary operations, increase throughput, reduce off-spec product generation, and improve yields.
- *Reduced regulatory burden.* Improving environmental performance and achieving performance goals that exceed compliance have been demonstrated in many P2 programs, which in turn reduce the costs of compliance.
- *Decreased liability.* Handling hazardous and toxic materials brings along with it high liabilities should an accident such as a fire, explosion, or a major spill occur.
- *Improved environmental health and safety.* P2 practices can be applied to all forms of pollution media. Reductions in pollution minimizes worker exposure and conserves resources and landfill space.

Many companies that claim to have pollution prevention programs really undertake P2 activities only occasionally. Such enterprises often point to the fact that they have a string of successes they attribute to recycling or perhaps true source reduction achievements. But these are not programs. They are only isolated opportunities that have been captured.

In many cases, P2 practices are based on recognizable opportunities with well-defined Tier 1 and 2 cost savings. A few successes often encourage companies to promote the spirit of P2, and indeed many environmental policy statements often incorporate a commitment to reducing pollution at the source. However, when one examines how P2 is implemented, one sees that the program is only a small element of the EMS and is rarely applied on a dedicated and continual basis. This creates two problems. First, the benefits of continual P2 improvements to environmental performance, such as cumulative efficiency gains, costs savings, and reductions of long-term liabilities can never really be achieved. Second, P2 activities are often implemented for the wrong reasons; hence, the most cost-effective strategies are never applied.

The proper way to think of a P2 program is that it is a systematic approach to pollution management through the reduction of as many of the cost components that lie within the four cost tiers. By comparing the total costs of different technology options that achieve comparable levels of compliance, the most cost-effective strategies become apparent. When a company analyzes these strategies from a cost-benefit standpoint, what it really does is develop an investment strategy. In many cases, going beyond minimum environmental compliance requirements leads to the most cost-effective pollution management strategy, because through that a company can capture Tier 3 and 4 cost savings.

Logical investment strategies can be planned only by properly applying life-cycle cost analysis. LCCA is an economic evaluation method in which all costs associated with a project or investment are considered from cradle to grave. The analysis considers all costs arising from owning, operating, maintaining, and ultimately disposing of a project to be potentially important to making a decision on whether an investment is worthwhile.

In the context of pollution prevention, a company seeks to evaluate whether or not a strategy is more cost effective in the long run compared to the status quo of managing its environmental aspects. LCCA is particularly well suited for evaluating the financial merits of a P2 investment, because it can be applied to any capital investment decision in which higher initial costs are traded for reduced future cost obligations. Since P2 is an approach that has a direct impact on future cost obligations, especially those associated with long-term liabilities (i.e., Tier 3 costs), LCCA yields a significantly better assessment of the long-term cost effectiveness than alternative economic methods that focus on only initial investment or operating costs in the short run.

In contrast, the most common alternative method of economic evaluation is the *payback method.* The payback method generally focuses on how quickly the *initial investment* can be recovered. As such, it is not a measure of *long-term* economic performance or profitability. The method typically ignores all costs and savings occurring after the point at which the payback for the capital investment is reached. It also does not differentiate between project alternatives having different lives and often uses an arbitrary payback threshold. The most serious shortcoming with this method is that it ignores the *time-value of money* (TVM) when comparing future streams of savings against the initial investment cost. LCCA stands in direct contrast to the payback method. Table 4–3 compares the two.

Three supplementary measures of economic performance that are complimentary to the LCCA method of investment evaluation: *net savings* (NS), the *savings-to-investment ratio* (SIR), and *adjusted*

Table 4–3 Comparison Between LCCA and Payback Method

Economic Indicator	*LCCA*	*Payback Method*
Considers long-term profitability	Yes	No
Considers costs after payback is reached	Yes	No
Considers savings after payback is reached	Yes	No
Takes into account the TVM	Yes	No
Differentiates between different project or investment lives	Yes	No

internal rate of return (AIRR). These measures are based on the same stream of costs and savings over the same study period.

NS can be used in place of the LCCA measure itself to determine the most cost-effective P2 alternative when evaluating two or more mutually exclusive alternatives. Within any group of mutually exclusive P2 project alternatives, the alternative with the lowest life-cycle cost (LCC) also has the highest NS.

The SIR and the AIRR measures are useful for ranking independent projects when faced with a budget insufficient to fund all the cost-effective project opportunities (both P2 and non-P2 projects) within a company or operating department. The SIR and AIRR should not be used to identify the most cost-effective P2 alternative, nor should they be used as a basis for selecting the most cost-effective project when evaluating P2 and non-P2 projects.

In defining and delineating the requirements of any economic analysis, it is helpful to identify the type of investment decision to be made. Five primary types of investment-related decisions for P2 are

1. Accept or reject a single P2 opportunity.
2. Select an optimal-efficiency-level P2 action.
3. Select an optimal P2 alternative from competing alternatives.
4. Select an optimal combination of interdependent P2 alternatives.
5. Rank competing P2 actions to allocate a limited budget.

Examples of each of these investment categories follow.

Accept or Reject a Single P2 Opportunity

Sour water from the refining section is sent to the sour water stripping system to remove H_2S, NH_3, and phenols. There are two sour water strippers. The major environmental aspect is the air emission problem (H_2S). The high concentration of H_2S and NH_3 in the generated wastewater from the strippers indicates low stripper efficiency. The refinery is considering an investment into a sulfur recovery unit. This is a high-cost P2 investment. Approximate costs for this investment are in excess of $7 million.

Select an Optimal-Efficiency-Level P2 Action

The refinery has six watertube boilers with a combined steam production capacity of 130 ton/h. By applying material and energy balances, the team identified significant losses attributed to boiler blowdown and excess condensate. As water evaporates in the boiler steam drum, solids or scale present in the feed water are left behind. The blowdown purpose is to limit the concentration of impurities in the boiler water. The right amount of blowdown is critical. Too much can result in energy losses and excessive chemical treatment cost, while too little will allow excessive concentrations of impurities to build up and damage tubes. Watertube or drum-type boilers have two blowdown systems: continuous blowdown from the steam drum to limit the concentration of the water, and intermittent blowing from the bottom headers to remove suspended solids from the boiler. During an assessment, the total boiler blowdown rate was found to be 6 m^3/h. This entire amount is drained to the refinery's wastewater treatment plant. Minimizing this blowdown rate reduces materials and energy losses. Since the temperature of the blowndown water is the same as that of the steam generated in the boiler, any savings in blowdown also reduces the makeup feedwater and the chemical treatment additives. The instability in the controlling process is obvious from the data, indicating that the blowdown system is ineffective. To calculate the boiler's concentration ratio, the silica concentration was used, since it is one of the critical controlling limits for watertube boilers. Silica comes form several types of deposits, which are considered very difficult to remove (e.g., amorphous silica). The amount of $240,900 is the total savings that could be achieved yearly by better control of boiler blowdown and better management of the operation. To capture the savings, the refinery should change the current blowdown approach and optimize the boiler's operating window as opposed to performing random blowdown. This eliminates operator error and provides system stability. An automatic blowdown control system optimizes surface blowdown by regulating the water volume discharged in relation to the amount of dissolved solids present, conductivity, and silica or chlorides concentrations. A probe provides feedback to a controller/

driving mechanism through a modulating blowdown valve. An alternative control configuration is a proportional control with the blowdown rate set proportional to makeup water flow. The approximate costs for an automated systems is $23,239. Total cost for six boilers is $140,844, with a payback period of seven months maximum.

Select an Optimal P2 Alternative from Competing Alternatives

The refinery's cooling towers consume approximately 120 m^3/h as makeup water from wells. About 45–50% of the cooling water is lost through evaporation, and the remaining percentage is discharged as blowdown with a conductivity of 5000–6100 $\mu s/cm^2$. The makeup water has relatively high salinity. The high TDS (total dissolved solids) content in the makeup water requires a high concentration ratio, increases water and chemical consumption, and therefore blowdown frequency. The refinery is considering two approaches to managing blowdown and cooling tower operational efficiency. One is to optimize chemical dosing, including automation of blowdown treatment operations. Another is to consider installing a reverse osmosis unit, which would feed the boilers and cooling towers with higher-quality water, thereby improving the overall performance of the cooling towers. While this entails a higher capital investment, water and chemical treatment demand reductions could make the investment attractive.

Select an Optimal Combination of Interdependent P2 Alternatives

To improve cooling tower operational efficiency and reduce blowdown and makeup water demand, the refinery is studying and comparing the advantages or disadvantages and potential savings from several options. Among these are

1. The refinery has two ion exchangers units to demineralize water that serves as feed to the boilers. Cycle time for both units is 4 to 5 days and NaOH and HCl are used for the resin

regeneration. The generated wastewater from the regeneration process is collected in a neutralizing pit and used for neutralizing the wastewater at the wastewater treatment plant, bypassing the API separator, since it has a pH of 2–3. Since the total volume of the generated water from the regeneration is larger than volume of the neutralizing pit, the backwash and a portion of the rinse water is sent to the southern pool, near the cooling towers. This pool is used to collect the sand filter's backwash, part of the regeneration effluent, and sometimes part of the cooling tower blowdown. The pool water also is diverted to the API separator in the wastewater treatment plant, which increases the amount of sludge generated from the plant to the sludge ponds. Since part of the regeneration effluent has good quality, it can be reused. The water generated from backwash and rinsing has low conductivity; therefore, it could be used as cooling tower makeup water in addition to the RO (reverse osmosis) water. However, the water generated from the rinsing step in the first 10 minutes have high conductivity, so it should be collected in the neutralizing pit with the rest of the regeneration water (regenerate injection and displacement stages). Only the low-salinity part of the regeneration effluent (backwash and rinsing after the first 10 minutes) should be collected in the southern pool. Since the volume of this pool is small (around 110 m^3) the cooling towers could be fed with part of this water during the regeneration operation.

2. A new RO unit will be installed for the heat-exchanger makeup water. This unit will have an impact on the operation of existing ion exchangers. Since the conductivity of the new RO is lower than an existing unit, this will lead to an increase in time between the ion exchange regenerations and decrease the generated wastewater from ion exchangers regeneration. This will result in a cost savings in water and chemicals usage. Over time (within two years), the effectiveness of the new RO will decrease. This will lead to an increase in the RO conductivity.

Rank Competing P2 Actions to Allocate a Limited Budget

Table 4–4 lists examples of opportunities identified during a P2 audit of a refinery and the ranking performed in terms of savings and payback periods.

For a company to stay on track and capture the maximum benefits of a P2 program, it must establish that program as its basis for continual improvement of environmental performance and religiously apply LCCA to assessing alternative environmental strategies.

4.6 Audits and Management Review

The reader is likely familiar with many types of audits. Common ones include environmental audits, financial audits, and P2 audits, on which my previous publications focus. Here, the discussion focuses on the EMS audit.

EMAS defines an audit as "a systematic, documented, periodic, and objective evaluation of the performance of an organization, management system, and processes designed to protect the environment." The objective of an EMS audit is to verify the organization's management system and environmental policy. In other words, the audit is designed to gather information to be used to determine if the company in fact is on the right track with regard to its EMS, and if not, then the audit will identify the deficiencies and point the way back to the right path. Such audits are not intended to review the EMS itself. That is the responsibility of the organization's top management. It is top management's responsibility to determine and establish what type of corrective actions are needed based on a management review using the information obtained from the audit and by considering recommendations made from the environmental management team and other resources from within the organization.

The value of the audit is that the information can be applied to help management control those work practices that have an effect on the environment and to gauge how well the company is complying with its own policies. It may also be used to determine the existence and proper functioning of the management system elements. It does not focus on whether specific performance complies with legal or

Table 4–4 Example of Ranking P2 Opportunities

Pollution Prevention Measure	*Materials Savings, ton/yr*	*Materials Savings, $/yr*	*Cost of Investment, $*	*First Year Saving, $*	*Payback, months*
Converting the boilers' manual blowdown to automatic blow-down-control systems (savings of water, energy, and chemicals)	24,090	240,900	139,437	101,463	7
Install control valve at service waterline, cascade with alarm high level of the low condensate drum	8,760	3,085	4,225	–1,141	17
Repair all steam traps and install a computerized steam trap management system to improve the overall maintenance program and reduce steam leakage	105,120	631,608	19,718	611,890	1
Reuse the RO flushing water and sand filter backwash water for cleaning and as fire water through collecting it in holding tank	37,047	13,045	21,217	–8,083	24
Reuse the water generated from the ion exchanger backwash and rinsing as makeup water for cooling towers	24,612	27,732	704	27,028	0.5
Provide residential area with flow meters	74,460	26,218	14,084	12,134	7
Provide caps on all hoses in the unloading section, and install hopper tank with pumps to recycle the drained crude oil	3,650	514,085	42,254	471,831	1

Source: Work performed under the direction of the author, USAID Contract No. LAG-I-98-00034-00, Task Order 816, CDM Project No. 3029-43173, *Task 3 Report on the Jordan Petroleum Oil Refinery*, Amman, Jordan, 2005.

other requirements (again, that is top management and the environmental manager's responsibilities). In addition to determining conformance to the EMS criteria, an EMS audit meets several other objectives: It determines whether the EMS is properly implemented and being maintained; it identifies areas of improving the EMS; it assesses the capabilities of the internal management review process to ensure continuing suitability and effectiveness of the EMS.

An EMS audit can be conducted internally or through the use of outside auditing firms. Either way the auditors must conduct the exercise on the basis of an independent, impartial, and objective assessment. If the audit is conducted internally, then the team should comprise members independent of the site being audited. For a single facility, this can mean conducting a series of audits on a department-by-department basis, with rotating teams from different divisions of the facility. There are pros and cons to both approaches. Since the audit focuses on the EMS itself, the importance of technical competency for individual refinery processes is less important, and hence a team comprising external auditors should be the most impartial of the two alternatives. However, the use of internal talent familiar with the products and processes does offer the possibility to identify opportunities that can not only improve the EMS but direct the attentions of the EMS management and P2 teams.

As in the case of any audit, the protocols, objectives, and scope should be clearly defined at the onset. The extent of the audit, its boundaries, specific activities, and the form in which results are to be reported should be defined at the start. The lead auditor should prepare a detailed list of activities and subjects to be covered, an implementation schedule should be established, and the program reviewed and approved by management prior to the start. All team members should participate in developing the protocols and overall scope and schedule.

Audit cycles (i.e., the frequency of audits) are somewhat subjective. Except for EMAS, none of the EMS standards specifies a rigorous frequency. EMAS calls for audits every three years, and as a general rule, this is what is practiced by companies that have formal ISO

14001 registration. The ultimate decision lies with management. The costs for audits are significant, especially when outside auditors are engaged. On the other hand, an ounce of prevention is worth a pound of cure. The right way to determine the optimum frequency of audit cycles is to first examine how diligent the organization is in conducting audits of individual operational aspects of the EMS itself and how effective these audits are. Significant environmentally-related activities, such as treatment of effluents, waste disposal practices, preventive maintenance inspections, SHE (safety, health, and environment) audits, security audits, and the like all require a certain diligence and high frequency of auditing to ensure operational performance. Their effectiveness should be examined on a frequent and regular basis and used in part as a basis to establish a frequency for the EMS audit.

Management has the responsibility to conduct a formal review of the audit findings. It is the role of management to assess the continuing suitability, adequacy, and effectiveness of the EMS. Such reviews need to be relatively broad in scope to address all activities, products, and services of the organization. This review should encompass financial performance and consider competitive position as well. Because such audits are extensive in nature, not all components of the EMS review can be assessed at once. Many aspects often require a collective review over a period of time.

In accordance with ISO 14004, the review process should include

- A review of environmental objectives and targets.
- The findings of the EMS audits.
- An evaluation of the suitability of the organization's environmental policy.

On the last point, the evaluation should consider possible changes or improvements based on consideration of changing legislation, changes in expectations and requirements of the organization itself, changes in products or activities within the organization, technology changes, lessons learned from environmental-related incidents, market forces and preferences, and both internal and external reporting and communication needs.

Because the intent of an EMS is to focus on continual improvement in environmental performance, it is quite possible that a critical and detailed review may result in changes in the environmental policy of the company, which in turn will cause the need for changes in the EMS itself.

4.7 Environmental Reporting and Communication

Environmental communications are both of an internal and external nature for an organization. The role of internal communication and its importance are obvious. Clear lines of ongoing communication are crucial from every aspect of a business, and it is especially important to top management for decision making purposes. Communication is also a key element in the documentation of environmental practices and provides legal evidence that the corporation has acted in a responsible manner. While there still remains a philosophy among some companies that act irresponsibly with regard to environmental management that files should be purged of "smoking guns," the point behind adopting and applying an EMS is that incriminating documentation of poor performance and practices cannot be created because the efforts are on the elimination of such practices.

It is also obvious that external communications are important to external stakeholders. External stakeholders include shareholders, legislators and regulators, employees, local communities, suppliers, customers and consumers, environmental groups including nongovernmental organizations (NGOs), financial analysts, lending institutions, insurers, industry trade groups, contractors and subcontactors, and others. Regular and clear communication with external stakeholders is crucial to attracting capital, maintaining a favorable public image and maintaining and growing the company's reputation, gaining market share, and attracting and retaining the best and most-talented employees, among other things.

Most corporations choose to disclose annual environmental reporting, coinciding with financial reporting. This makes perfect sense since the environmental and financial bottom lines of a corporation are inseparable in today's business world. In terms of the environmental standards, neither ISO 14001 nor BS 7750 requires a corporate environmental report. EMAS requires a comprehensive external environmental report. But, again, the practice of providing a yearly corporate environmental report as a part of the financial reporting of a corporation is so widespread today that requirements are not needed.

A comprehensive external environmental report generally covers the following topics:

- Corporate environmental policy.
- Environmental strategy.
- Description of the EMS components.
- Policy regarding environmental aspects that relate to products and services.
- Quantified listing of all inputs (material and energy) and outputs (air emissions, solid wastes, wastewater effluent) over a period of time.
- Assessment of compliance status.
- Evaluation of overall environmental performance.
- Description of the environmental management programs, including objectives, targets, measures, and schedules.
- Highlights of P2 achievements.
- Relationships with external stakeholders.
- Audit report findings and key recommendations.

Communication is not only a yearly report, it must become a strategy that goes beyond issuing a document. That strategy should take into account the specific interests of major financial stakeholders like shareholders, banks, insurers, and financial analysts. It behooves management to demonstrate that the chosen environmental strategy is efficient, effective, and providing financial returns to the shareholders. If the environmental strategy does not create a future cash flow or increase the value of assets, then it is simply economically unsustainable.

Recommended References and Websites

Websites

http://www.epa.gov/ems/. This is the U.S. Environmental Protection Agency website on environmental management systems with links to multiple sites.

http://ndems.cas.unc.edu/. This website presents the research results of the National Database on Environmental Management Systems, funded by the U.S. Environmental Protection Agency.

http://www.iso14000-iso14001-environmental-management.com/14000.htm. This website provides an ISO 14000 toolkit.

http://www.epa.state.oh.us/opp/ems/emsinfo.html. The Ohio Environmental Protection Agency website contains basic information, links, and general toolkits for EMS and P2.

http://www.ansi.org/conformity_assessment/accreditation_programs/accreditation_ems.aspx?menuid=4. American National Standards Institute's website. The American National Standards Institute (ANSI) and the American Society for Quality (ASQ) established the ANSI-ASQ National Accreditation Board (ANAB) for registrars of environmental management systems to ensure that the marketplace, including buyers, sellers, public agencies, and public interest organizations, can rely on the competence of ANAB-accredited bodies for their activities related to environmental management systems. This site describes their program and services.

http://www.uneptie.org/outreach/business/ems.htm. This is the United Nations Environment Program website. The site provides training materials and toolkits for EMS implementation.

http://www.epa.gov/Region3/ems/emspubs.htm. This is a U.S. EPA website. It contains the following publications: *An Implementation Guide for Small- and Medium-Sized Organizations; How to Implement an EMS PEER Center; An Environmental Management System Troubleshooters' Guide for Local Governments; EMS: A Better Bottom Line for Local Government; EMS Integration Project; Environmental Management System Pilot Program for Local Government Entities; Standardizing Excellence: Working with Smaller Businesses to Implement Environmental Management Systems; Environmental Management Systems Primer for Federal Facilities; Design Your Environmental Management System Workshop; Understanding and Implementing an Environmental Management System; IEMS Implementation Guide: Integrated Environmental Management Systems (IEMS)—Design for the Environment; Moving Toward Comprehensive Utility Management Systems: Report of the Environmental Management Systems (EMS) for Public Utilities Integration Project; Biosolids EMS Guidance Manual* (Updated 6-18-2002); *Environmental Management Systems Tools: A Reference Guide; The Small Business Source Book on Environmental Auditing; Third-Party Auditing of Environmental Management Systems: U.S. Registration Practices for ISO 14001; EMS Case Study: City of Gastonia Wastewater Treatment Division.*

Books

The following books can be obtained through Amazon (www.amazon.com). These volumes have average customer ratings of between four and five stars. Most of these works have been relied on in my own reviews of the subject matter. The full citations are not provided as the reader can obtain all details from the Amazon website.

Bendavid-Val, A., and N. P. Cheremisinoff. *Achieving Environmental Excellence: Integrating P2 and EMS to Increase Profits* (hardcover, February 28, 2003).

Cheremisinoff, N. P., and A. Bendavid-Val. *Green Profits: The Manager's Handbook for ISO 14001 and Pollution Prevention* (hardcover, April 16, 2001).

Edwards, A. J. ISO 14001 *Environmental Certification Step by Step*, rev. ed. (paperback, January 8, 2004).

Jackson, Suzan L. *The ISO 14001 Implementation Guide: Creating an Integrated Management System.* Wiley Series in Environmental Quality Management (hardcover, February 20, 1997).

Kanholm, Jack. *ISO 14001 in Our Company, Self-Study Course for Personnel* (paperback, July 1, 1998).

Kinsella, John, and Annette D. McCully. *Handbook for Implementing an ISO 14001 Environmental Management System: A Practical Approach* (paperback, November 2003).

Parry, Pamela. *The Bottom Line: How to Build a Business Case for ISO 14001* (paperback, November 23, 1999).

Schoffman, Alan, and Allan M. Tordini. *ISO 14001: A Practical Approach.* American Chemical Society Publication (hardcover, April 13, 2000).

5

Tools for Conducting the Initial Environmental Review

Introduction

Two software products accompany this book. Both can assist in organizing the information from an IER and performing an assessment that will enable assigning priorities to the environmental aspects identified.

As explained earlier, these are among the first important steps to implementing an EMS. By establishing a baseline characterization of how the facility manages its regulatory obligations and other environmental aspects, areas of nonconformance with the regulations and poor performance leading to negative impacts can be identified. These deficiencies, or so-called gaps, can be filled in by devising corrective actions, such as pollution prevention programs or upgrading pollution controls. Because of the nature of manufacturing operations, and particularly refineries, a large number of negative environmental aspects are likely to have been marginally managed or not managed at all. This is true even if the facility has had a reasonably good compliance record. It is important to keep in mind that the absence of fines, penalties, and citations over a period of years unto itself is not assurance that a facility has exemplary environmental performance. My experience has been that this simply reflects that the facility has been operating under the radar screen of the regulator and long-term pollution problems have gone unchecked. These will surface eventually and create significant financial threats to an organization. Furthermore, compliance unto itself should not be the focus of attention. Pollution is waste, and waste is inefficiency. Inefficiency means higher operating costs and reduced profits. So, the real benefit of an IER is to identify where a company is loosing money from poor performance and implement actions that reduce those losses and, in a number of instances, future threats that could affect its sustainability and growth.

To manage a significant number of negative environmental aspects, the company must devise not only specific corrective actions but environmental action plans to implement the corrective actions. It also must devise an overall strategy for tackling these over time. The reader should agree that it is logical to address the highest priorities first, then systematically work through the list of negative environmental aspects identified as requiring action. To accomplishing this, after the aspects have been identified, they should be ranked according to a systematic set of criteria. Both criteria for ranking the aspects and software to facilitate the task are included in a second software tool discussed in this chapter and provided with the book's CD-ROM.

At the risk of being redundant, I want to emphasize that the focus of an EMS should not be reducing emissions and focusing on meeting statutory requirements. Chapter 4 tools (the environmental effects register and environmental regulations register) help identify and bring regulatory obligations into perspective, as they are legal requirements to continue baseline business operations. But, meeting compliance obligations alone will not maximize profitability. Maximum profits can be gained only by reducing the wastes and inefficiencies within the manufacturing steps, services, and even products that the company markets and distributes. Hence, the proper approach to conducting an IER should not focus on compliance per se.

Conformance to the EMS requirement of meeting environmental compliance obligations is achieved naturally through improved environmental performance. By systematically reducing wastes, inefficiencies, and pollution at the sources of generation, emissions become less. Lower emissions translate into overall improved environmental performance and a lowering of operating costs, because waste and pollution management are simplified. Hence, the facility's compliance obligations not only are met but exceeded. As explained in my prior publication *Achieving Environmental Excellence: Integrating P2 and EMS to Increase Profits* (Rockville, MD: ABS Consulting/Government Institutes, 2003), the necessarily similar structures of EMSs and P2 programs make it possible to integrate the elements of a P2 program smoothly into an EMS. When this is done, P2 is raised from being a mere commitment within the EMS to being a major operational component of the EMS. Instead of looking just at the regulatory compliance status and environmental aspects of the company's operations, then at ways to improve on these, you also specifically and systematically look for high-return P2 opportunities from the outset, and do so continuously. An integrated EMS/P2 program reduces operating costs and the risks of environmental liability more, better, and faster than a traditional EMS. The earlier major cost savings can be used to help finance investments, further accelerating improvements in business and the environmental performance of the company.

The software developed with this publication reflects this approach. The software provides a platform for data collection by the IER audit team, where the type of information collected enables the team members to review and assess P2 opportunities during the analysis stage.

The chapter provides an explanation on how to use the two software products. There are two programs: IER Auditor and EA (Environmental Aspects) Evaluator. IER Auditor is an audit questionnaire to assist in identifying regulatory requirements (from a U.S. perspective), environmental aspects, and P2 opportunities. It is a database management tool that runs on Microsoft Access. Your PC requires Microsoft Access to operate both programs. IER Auditor helps create a baseline report that can be used by the audit team to identify areas of nonconformance and poor environmental performance. From the baseline, opportunities for improvement become apparent.

EA Evaluator enables the user to perform a simplified analysis to rank EAs. This provides a starting base for developing environmental action and management plans and an overall strategy.

For those readers not intimate with the complex environmental regulations within the United States, an overview of these is first provided. Terms used in IER Auditor as part of the questionnaire rely on terminology and definitions used in the regulations.

Both programs are contained on a single CD. When you open the CD you will see two files: IER Auditor Software Package and EA Evaluator Software Package. Drag each of the folders into the directory of your choice. Read over Sections 5.2 and 5.3 for more information on getting started.

5.1 U.S. Regulations, Waste Classifications, and SIC Codes

5.1.1 U.S. Environmental Laws and Regulations

Before describing the software tools, a brief summary of major U.S. environmental regulations is provided. The regulations appear as part of the audit questionnaire in IER Auditor. IER Auditor is written for a generalized audit with a focus on identifying environmental aspects of significance; however, a number of U.S. terms with regard to waste classification and environmental statutes are included among the audit questionnaire sections. Non-U.S. readers may wish to skip over this discussion and go directly to the next section.

It is important to distinguish between laws and regulations. Congress, and only Congress, enacts federal laws. Federal executive departments and administrative agencies write regulations to implement the authority of laws. Regulations (as well as executive orders and proclamations) are ancillary or subordinate to laws, but both laws and regulations are enforceable. The *U.S. Code* is the official compilation of codified laws by subject, the *U.S. Statutes-at-Large* is the official chronologic compilation of all laws, and the *Code of Federal Regulations* is the official compilation of regulations.

The law or "act" is compiled annually in the *U.S. Statutes-at-Large* (Stat.). The *United States Statutes-at-Large* contains all laws, concurrent resolutions, and joint resolutions enacted during a session of Congress. It also includes reorganization plans, proposed and ratified amendments to the Constitution, proclamations by the president, and legislative histories. It also contains laws that apply to the general populace. These are known as *public laws* (PL).

The act is then codified in the *U.S. Code* (U.S.C.), which is published every six years with annual supplements. The *U.S. Code* codifies by category only public laws of general and permanent nature in effect on the date of publication. The vast quantity of expired and repealed laws since 1789 is not included in the U.S.C.

The *U.S. Code* is most often used to find the current "laws of the land." Aside from compiling laws by subject matter, the U.S.C. contains the history of recent amendments and cross-references and other notes of pertinence to aid the researcher. The *U.S. Code* also contains the Index of Acts by Popular Name, General Index, and several tables.

Although the statutes are the legal reference, 19 titles of the *U.S. Code* have been enacted into law and therefore supersede the specific statutes from which they were enacted.

The annual, cumulative supplements to the U.S.C. contain all changes in or additions to the general and permanent laws and *U.S. Code* ancillaries since the last edition of the U.S.C.

There are several ways to find laws in the *U.S. Code*. The Index of Acts by Popular Name lists the names of laws by their popular designation or short title, the date of passage, the public law number, its location in the *Statutes-at-Large* volumes, and the *U.S. Code* title number and sections. A history of amendments to the original law is listed, although it may be incomplete.

The *Federal Register* consists of two publications, the annually revised *Code of Federal Regulations* (CFR) and the daily *Federal Register* (FR). These two publications should be used together to find the latest version of a regulation.

The *Code of Federal Regulations* codifies the general and permanent rules of executive departments and agencies that have been published in the *Federal Register*. The CFR is divided into 50 titles, which represent broad areas subject to federal regulation. Each title is divided into chapters, which usually bear the name of the issuing agency. Each chapter is further subdivided into parts covering specific regulatory areas. Throughout this guide, the CFR is cited by title and part, followed by the part's name and the agency promulgating the regulation.

The *Federal Register* is the daily supplement to the CFR. It serves as the mechanism to provide official notification to the public about federal documents or proposals having general applicability. These may include presidential proclamations and executive orders and federal agency rules, regulations, and notices. The FR also serves to notify the public and interested parties about an agency's intent to prepare an environmental impact statement or proposed changes to an agency's rules and regulations. It further provides the mechanism for obtaining comments from affected parties regarding the proposed action by the agency.

The list of CFR sections affected (LSA) serves as an important tool for determining the current text of a specific CFR title and section and for finding changes made to the CFR section. The LSA may be used as a checklist to determine how particular CFR sections might have been affected by subsequent rule changes. The LSA lists current regulations as well as regulations that were revoked, terminated, or superseded. Current and proposed and rules and regulations are also listed, along with their status. The LSA is included in each FR issue and monthly changes are also published.

Among the environmental laws enacted by Congress through which the EPA carries out its efforts are

1938 Federal Food, Drug, and Cosmetic Act

1947 Federal Insecticide, Fungicide, and Rodenticide Act

1948 Federal Water Pollution Control Act (also known as the Clean Water Act)

1955 Clean Air Act

1965 Shoreline Erosion Protection Act

1965 Solid Waste Disposal Act

1970 National Environmental Policy Act
1970 Pollution Prevention Packaging Act
1970 Resource Recovery Act
1971 Lead-Based Paint Poisoning Prevention Act
1972 Coastal Zone Management Act
1972 Marine Protection, Research, and Sanctuaries Act
1972 Ocean Dumping Act
1973 Endangered Species Act
1974 Safe Drinking Water Act
1974 Shoreline Erosion Control Demonstration Act
1975 Hazardous Materials Transportation Act
1976 Resource Conservation and Recovery Act
1976 Toxic Substances Control Act
1977 Surface Mining Control and Reclamation Act
1978 Uranium Mill-Tailings Radiation Control Act
1980 Asbestos School Hazard Detection and Control Act
1980 Comprehensive Environmental Response, Compensation, and Liability Act
1982 Nuclear Waste Policy Act
1984 Asbestos School Hazard Abatement Act
1986 Asbestos Hazard Emergency Response Act
1986 Emergency Planning and Community Right-to-Know Act
1988 Indoor Radon Abatement Act
1988 Lead Contamination Control Act
1988 Medical Waste Tracking Act
1988 Ocean Dumping Ban Act
1988 Shore Protection Act
1990 National Environmental Education Act

A brief description of some of the major laws follows.

Superfund Reauthorization Act (1986)

The "Superfund" legislation provides for the cleanup of abandoned hazardous waste sites as well as contaminating accidents, spills, and other pollution-related emergencies. The EPA was given power to prosecute the companies responsible for the contamination and force them to assist in the cleanup. The Superfund also pays for cleaning sites for which no responsible party can be identified or who no longer exist.

Emergency Planning and Community Right-to-Know Act (EPCRA)

Title III of the Superfund Amendments and Reauthorization Act (1986), EPCRA is intended to improve local community access to information about chemical hazards and to improve state and local emergency response capabilities. State right-to-know acts also require reporting of various kinds of chemical hazard information. A description can be found in Chapter 4.

Comprehensive Environmental Response, Compensation, and Liability Act (CERCLA)

Chapter 103 of 42 U.S.C., Section 9601 et seq. (1980), addresses cleanup of hazardous substances. It empowers the EPA to identify and set priorities on sites for cleanup and order or carry out environmental remediation. Subject to limited defenses, it imposes strict liability for environmental cleanup on persons whose actions caused the release into the environment. It also mandates reporting hazardous substance releases to the National Response Center. In conjunction with the Clean Water Act, it mandates preparation of the national contingency plan for responding to oil or hazardous substances release.

Endangered Species Act (ESA)

Chapter 35 of 16 U.S.C., Section 460 et seq. (1973), provides for the listing of endangered and threatened species of plants and animals and designation of critical habitats for animal species. It establishes federal policy that federal agencies, in exercise of their authority, shall seek to conserve endangered species. It prohibits federal agencies from taking any action that would adversely affect any endangered or threatened species or critical habitat. The list of endangered and threatened wildlife and plants can be found at 50 CFR Part 17 (see later).

It also establishes a consultation process involving federal agencies generally and federal wildlife management agencies to facilitate avoidance of agency action that would adversely affect species or habitats. It prohibits all persons subject to U.S. jurisdiction, including federal agencies, from "taking" endangered species. The taking prohibition includes any harm or harassment, including habitat modification, and applies within the United States and on the high seas. The legal definition of *harm* can be found at 50 CFR Part 222.

National Environmental Policy Act of 1969 (NEPA)

Chapter 55 of 42 U.S.C., Section 4321 et seq. (1969), mandates federal agency consideration and documentation of environmental impacts of proposed actions and legislation. It also mandates preparation of comprehensive environmental impact statements where proposed action is "major" and significantly affects the quality of the human environment.

Oil Pollution Control Act of 1990

Chapter 40 of 33 U.S.C., Sections 2702 to 2761, mandates extensive planning for oil spills from tank vessels and onshore and offshore facilities. It establishes comprehensive elements of damage for oil spills and disposes strict liability on those responsible for oil spills but is not applicable to public vessels.

Pollution Prevention Act

Chapter 133 of 42 U.S.C., Sections 13101 and 13102, and Sections 6602 et seq. (1990), establishes the national policy that "pollution should be prevented or reduced at the source whenever feasible. Pollution that cannot be prevented should be recycled in an environmentally-safe manner. Disposal or other release of pollutants into the environment should be employed only as a last resort and should be conducted in an environmentally sound manner."

Resources Conservation and Recovery Act (RCRA)

Chapter 82 of 42 U.S.C., Section 321 et seq. (1976), addresses hazardous-waste management. The Resource Conservation and Recovery Act of 1976 amended the Solid Waste Disposal Act. It established a regulatory structure for the management of solid and hazardous wastes. Subtitle C of RCRA addresses "cradle-to-grave" requirements for hazardous waste from the point of generation to disposal. Subtitle D of RCRA contains less-restrictive requirements for nonhazardous solid waste. The hazardous and solid waste amendments (HSWA) of 1984 established additional waste management requirements and added Subtitle I, which imposes management requirements for underground storage tanks (USTs) that contain petroleum or hazardous substances. Most RCRA requirements are not industry specific but apply to any company that transports, treats, stores, or disposes of hazardous waste.

Regulations implementing Subtitle C of RCRA for hazardous-waste management appear in 40 CFR 260-279. Although RCRA is a federal statute, many states implement the RCRA program. In addition, many states have state-level hazardous-waste requirements that go above and beyond the federal RCRA requirements. Sites that have been contaminated with hazardous waste may be also subject to requirements under the Comprehensive Environmental Response, Compensation, and Liability Act, also commonly known as *Superfund.*

For a hazardous material to be regulated as a hazardous waste, it must first fall under the regulatory definition of solid waste, then within the definition of hazardous waste, both of which are described in 40 CFR 261, Identification and Listing of Hazardous Wastes. A waste can be hazardous either by exhibiting a characteristic of a hazardous waste or by falling under a category of listed hazardous wastes. Hazardous waste characteristics include toxicity, corrosivity, ignitability, and reactivity and are designated with the code *D*. Listed hazardous wastes include discarded commercial chemical products (designated with the code *P* or *U*), hazardous wastes from specific industries or sources (designated with the code *K*), and hazardous wastes from nonspecific sources (designated with the code *F*). Examples of each of these waste types are D001, ignitable wastes; U159, methyl ethyl ketone; K050, oily sludge from a petroleum refinery; and F001, halogenated solvents.

The foundation of RCRA is the manifest system. The manifest is a data sheet that identifies each waste shipment. The manifest accompanies the waste from the generating facility to the final disposal site and allows for cradle-to-grave tracking of the waste. Regulated entities that generate hazardous waste are subject to waste accumulation, manifesting, and record keeping standards. These regulations are found in 40 CFR 262 and include

- Becoming knowledgeable of applicable regulations.
- Training appropriate personnel about these requirements.
- Identifying all sources of hazardous waste and obtaining a facility ID number, if necessary.
- Establishing a waste minimization program to reduce or eliminate hazardous wastes.
- Appropriately packaging, labeling, and manifesting all hazardous waste generated.
- Meeting standards for hazardous waste accumulation.
- Complying with record keeping and reporting requirements.

The amount of hazardous waste generated determines a facility's generator category, which is used to establish the level of regulation. Facilities that generate less than 100 kg per month are conditionally exempt small-quantity generators (CESQGs) and are exempt from most reporting requirements, although they are still responsible for safe management of their waste. Facilities that generate between 100 and 1000 kg (220 and 2.200 lbs) per month are considered small-quantity generators (SQGs); facilities that generate over 1000 kg (2200 lbs) per month are considered large-quantity generators (LQGs).

Although long-term storage of hazardous waste requires a permit, generators can accumulate hazardous waste for shorter amounts of time without obtaining a permit. Accumulation limits are 90 days for LQGs and 180 days for SQGs.

Facilities that treat, store, or dispose of hazardous waste must obtain a permit, from either the EPA or a state agency that EPA has authorized to implement the permitting program. Subtitle C permits contain general facility standards, such as contingency plans, emergency procedures, record keeping and reporting requirements, financial assurance mechanisms, and unit-specific standards. Standards for hazardous waste treatment, storage, and disposal facilities can be found in 40 CFR 264 and 40 CFR 265. RCRA also contains provisions (40 CFR Part 264 Subpart S and §264.10) for conducting corrective actions that govern the cleanup of releases of hazardous waste or constituents from solid waste management units at RCRA-regulated facilities.

Based on the definition of hazardous waste, many recycling and reclamation activities involving hazardous waste are considered to be "treatment," depending on the particular recycling activities involved and the materials being recycled. The regulations covering recycling and reclamation of hazardous waste is complex. It is important for facilities to determine whether any processing of secondary materials is hazardous waste treatment that requires a TSD (transport, storage, and disposal) permit.

An underground storage tank includes the tank and any underground piping connected to it that has at least 10% of its combined volume underground. USTs containing petroleum and certain hazardous substances are regulated under Subtitle I of RCRA. The UST Program is administered in large part by state and local agencies, which may impose their own additional requirements. Subtitle I regulations (40 CFR Part 280) contain tank design and release detection requirements (requirements such as spill, overfill, and corrosion protection and regular leak detection, monitoring, and reporting), as well as corrective action (i.e., cleanup) standards for USTs.

Hazardous and Solid Waste Amendments (HSWA; 1984)

The amendments expanded the scope and increased the requirements of RCRA. HSWA addressed congressional concern about the adequacy of existing requirements to prevent uncontrolled releases of hazardous constituents or hazardous wastes from hazardous waste management units.

Water Pollution Prevention and Control (Federal Water Pollution Control Act, Clean Water Act)

Chapter 26 of 33 U.S.C. (1977) addresses water pollution. This act establishes the National Pollution-Discharge Elimination System (NPDES) permitting program, to control the discharge of pollutants from point sources into navigable waters. NPDES permits must incorporate industry-specific technology based on effluent standards. It also establishes the Dredge and Fill Permit Program, to control the discharge of dredged or fill material into navigable waters. It requires federal agencies to accommodate concerns of states regarding the consistency of federal projects with state non-point source pollution control programs and bans discharge of pollutants into navigable waters.

In addition to these environmental regulations, an important safety standard under the Occupational Safety and Health Administration is the Hazard Communication Standard (HCS). The reader will find a discussion of the HCS in Chapter 4.

5.1.2 Waste Classifications

A hazardous waste is one listed as such by the U.S. Environmental Protection Agency or that exhibits one or more hazardous characteristics (also as specified by the EPA). Hazardous wastes are threatening to human health and the environment. Waste that displays one or more of four hazardous characteristics is a regulated hazardous waste:

1. *Ignitability.* Easily flammable, such as solvents.
2. *Reactivity.* Capable of rapid chemical reaction, such as peroxides.
3. *Corrosivity.* Highly acidic or alkaline, able to dissolve metals or burn the skin, such as hydrochloric acid or sodium hydroxide.
4. *Toxicity.* Capable of releasing toxic constituents into the environment, such as lead-based paint.

Industrial waste is any industrial waste not listed as hazardous and having little or no hazardous characteristics. Class 1 nonhazardous industrial waste can include certain levels of constituents and specified properties that, at higher levels, might otherwise render the waste hazardous.

Industrial wastes result from (or are incidental to) operations of industry, manufacturing, mining, or agriculture; for example, wastes from power generation plants, manufacturing facilities, and laboratories serving an industry. *Nonindustrial wastes*, by contrast, come from sources such as schools, hospitals, churches, dry cleaners, most service stations, and laboratories serving the public.

Nonhazardous industrial waste is referred to as *Class 1* waste. It is considered potentially threatening to human health and the environment if not properly managed, because of the constituents and properties this class can include. Therefore, there are special handling requirements for Class 1 wastes. An example is water contaminated with ethylene glycol. Examples of *Class 2* wastes include waste and activated sludge from biological wastewater treatment. *Class 3* includes materials such as demolition debris (e.g., bricks) that are insoluble, do not react with other materials, and do not decompose. Class 2 and 3 wastes are often accepted by local landfills. However, a Class 2 or 3 designation does not mean that the waste is incapable of causing harm in every management (or mismanagement) situation.

Listed hazardous waste are those on an EPA list, some 400 hazardous wastes. Descriptions of listed waste are found in 40 CFR Part 261, Subpart D, Sections 261.31–33. These wastes are often referred to as follows:

- F-listed waste is waste from nonspecific sources, Section 261.31.
- K-listed waste is waste from specific sources, Section 261.32.
- P-listed waste is unused acutely hazardous off-specification materials as well as container residues and spill residues of these materials, Section 261.33.
- U-listed waste is unused toxic hazardous off-specification materials as well as container residues and spill residues of these materials, Section 261.33.

5.1.3 SIC Codes

SIC stands for the Standard Industrial Classification code established by the Bureau of Census to track the flow of goods and services within the economy. The SIC code for the petroleum refining and related industries is 29.

The petroleum refining industry is classified as SIC 2911, which includes the production of petroleum products through distillation and fractionation of crude oil, the redistillation of unfinished petroleum derivatives, cracking, or other processes. Related industries that fall under SIC 29 are

- SIC 2951, Asphalt Paving Mixtures and Blocks
- SIC 2952, Asphalt Felts and Coatings
- SIC 2992, Lubricating Oils and Greases
- SIC 2999, Petroleum and Coal Products

Certain products that are manufactured by the petroleum refining industry are produced by other industries as well. These include the following:

- SIC 2865, Cyclic Organic Crudes and Intermediates and Organic Dyes and Pigments
- SIC 2869, Industrial Organic Chemicals
- SIC 2819, Industrial Inorganic Chemicals
- SIC 2821, Plastic Materials, Synthetic Resins, and Nonvulcanizable Elastomers
- SIC 2873, Nitrogenous Fertilizers
- SIC 4613, Refined Petroleum Pipelines
- SIC 5171, Petroleum Bulk Stations and Terminals

5.2 IER Auditor

IER (Initial Environmental Review) Auditor is a multipage audit questionnaire that is prepared as a simple database management tool using Microsoft Access as the platform. There are no special requirements to use the program other than Microsoft Access installed on a PC. Place the CD-ROM contained in the jacket of this book into the CD-ROM drive of the computer. Drag the file folder named IER Auditor Software Package onto the desktop or any subdirectory. Then double-click or right-click to open the folder. Click on IER Auditor to open the program.

When the program first appears you will see a security warning that states, "This file may not be safe if it contains code that was intended to harm your computer. Do you want to open this file or cancel the operation." You may ignore this warning and click on Open to make the program functional.

IER Auditor has multiple pages or spreadsheets. The first page is an index, which will bring you to various input pages (see Figure 5–1). The software is hyperlinked to a Tutorial that I recommend you read over to understand the operational features better. The following is a general description of the features. Note that Microsoft Access built security codes into the system, which this software product does not bypass. When you click on to the Tutorial link, a message will appear: "Hyperlinks can be harmful to your computer and data. To protect your computer, click only those hyperlinks from trusted sources. Do you want to continue?" Click on Yes and the Tutorial will open.

The first data entry spreadsheet is the Facility General Information file. You may enter information on the date of the IER audit, name of the facility, location, contact information, and other pertinent information on the subject facility using this subfile. At the bottom of the file is a Start button. Click on Start to go to the main index page of files.

There are 11 files or spreadsheets on which to gather. Clicking on the arrow next to each file title brings you to the data entry page for that subject.

File 1 is titled Environmental Policy Checklist. There are 21 statements or features that characterize a good corporate environmental policy statement. The features were identified from a review of more than 50 published corporate environmental policy statements, which are believed to reflect strong commitments to environmental

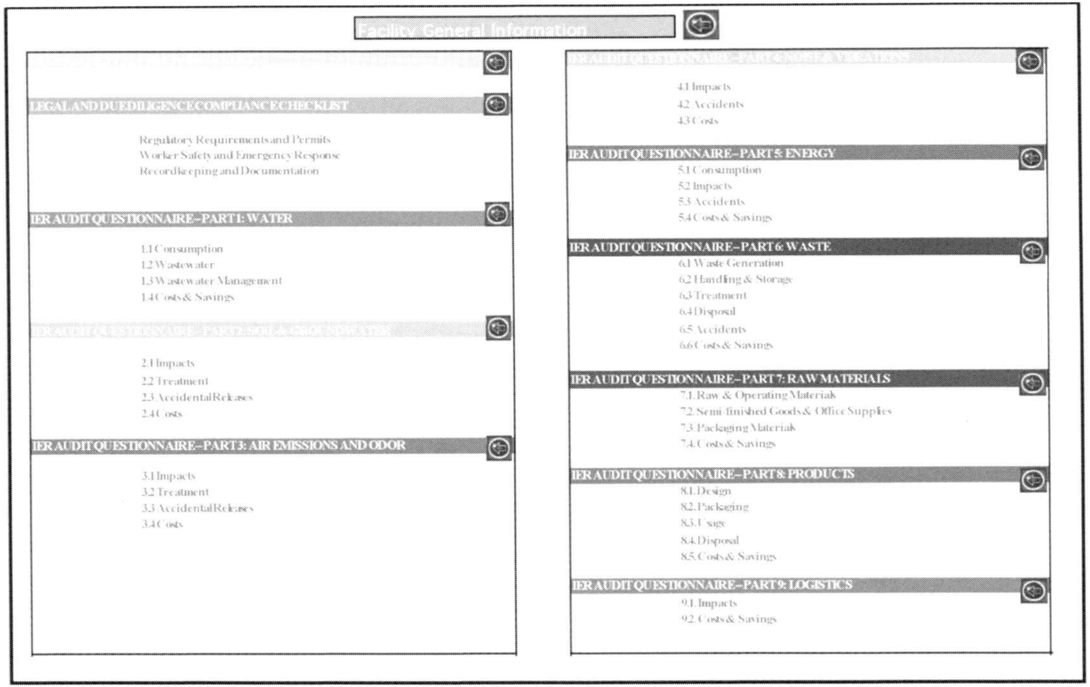

Figure 5-1 Index page to IER Auditor.

protection, public and worker safety, pollution prevention, and the principles of an EMS. You may compare your company's policy against each of the features by assigning a numerical rating from 1 to 5. The lowest value, 1, denotes that a company's policy does not embody the feature and the highest value of 5 denotes that the policy is a good match to the statement. The Environmental Policy Checklist sums the values you enter and provides a total absolute score and a percentage match. A perfect score is 100% based on a total of 160 points. You may save the results for future reference, print the page, or clear the data entries at the end of the exercise. Click on the Back button to return to the File Index page.

File 2 is titled Legal and Due Diligence Compliance Checklist. This file contains a list of questions that focus on regulatory compliance obligations and identifying available records from within the facility. There are three categories of information: Regulatory Requirements and Permits, Worker Safety and Emergency Response, and Recordkeeping and Documentation. Answer each question by clicking on Y for yes, N for no, and NA for not applicable.

File 3 is titled IER Audit Questionnaire—Part 1: Water. Four data collection categories enable you to document information from the audit: Consumption (i.e., water demand), Wastewater, Wastewater Management, and Costs and Savings.

File 4 is titled IER Audit Questionnaire—Part 2: Soil and Groundwater. Four data collection categories enable you to document information from the audit: Impacts, Treatment, Accidental Releases, and Costs.

File 5 is titled IER Audit Questionnaire—Part 3: Atmospheric Emissions and Odors. Four data collection categories enable you to document information from the audit: Impacts, Treatment, Accidental Releases, and Costs.

File 6 is titled IER Audit Questionnaire—Part 4: Noise and Vibrations. Three data collection categories enable you to document information from the audit: Impacts, Accidents, and Costs.

File 7 is titled IER Audit Questionnaire—Part 5: Energy. Four data collection categories enable you to document information from the audit: Consumption, Impacts, Accidents, and Costs and Savings.

File 8 is titled IER Audit Questionnaire—Part 6: Waste. Six data collection categories enable you to document information from the audit: Waste Generation, Handling and Storage, Treatment, Disposal, Accidents, and Costs and Savings.

File 9 is titled IER Audit Questionnaire—Part 7: Raw Materials. Four data collection categories enable you to document information from the audit: Raw and Operating Materials, Semi-finished Goods and Office Supplies, Packaging Materials, and Costs and Savings.

File 10 is titled IER Audit Questionnaire—Part 8: Products. Five data collection categories enable you to document information from the audit: Design, Packaging, Usage, Disposal, and Cost Savings.

Finally, File 11 is titled IER Audit Questionnaire—Part 9: Logistics. Two data collection categories enable you to document information from the audit: Impacts and Costs and Savings.

Once the information is collected and collated, the audit team may begin developing a baseline characterization of the facility. The auditors will be able to identify areas of noncompliance as well as parts of the business where significant negative environment impacts are likely to exist. The auditors should be able to begin developing environmental strategies for corrective actions based on a review of the information acquired from the audit.

Once the information has been documented and general strategies developed, the environmental manager, along with members of the IER audit team, can begin acquiring more-detailed information on specific environmental aspects and their impacts. Using EA Evaluator, it will be possible to develop a list of priorities of the significant environmental aspects. After this has been accomplished, effort can get underway to develop environmental action plans aimed at corrective actions.

5.3 Environmental Aspects (EA) Evaluator

EA Evaluator runs on a Microsoft Access platform. You require Microsoft Access on your PC to use the program. The intent of the software is to assist the environmental manager in developing a priority rank of environmental aspects. This enables strategy recommendations to be devised and corrective actions scheduled.

EA Evaluator contains a tutorial that explains and illustrates the operational features of the program. When opening EA Evaluator, go directly to the Tutorial page and read through the step-by-step instructions. To open the program, place the CD-ROM contained in the jacket of this book into the CD-ROM drive of a computer. Drag the file EA Evaluator Software Package (i.e., the entire folder) onto the desktop or any subdirectory. Then double-click or right-click to open the file.

Double-click on EA Evaluator. When the program first appears, you will see a security warning which states, "This file may not be safe if it contains code that was intended to harm your computer. Do you want to open this file or cancel the operation." You may ignore this warning and click on Open to make the program functional.

The software is hyperlinked to a tutorial that I recommend you read to understand the operational features better. When you click on the Tutorial link a message will appear: "Hyperlinks can be harmful to your computer and data. To protect your computer, click only those hyperlinks from trusted sources. Do you want to continue?" Click on Yes and the Tutorial will open.

To have a full appreciation of the analysis performed by EA Evaluator, it is recommended that you read this section, which explains the methodology and criteria used to set priorities on the environmental aspects.

The recommended methodology for identifying and ranking significant environmental aspects is a five-step method that can be laid out on to an Excel spreadsheet. Figure 5–2 illustrates the approach.

Step 1 is to note the name of the auditors and the date of the assessment. While this seems obvious, it is no less important than the remaining steps. Who performs the assessment and when it was done are important pieces of historical significance to the company. What comes out of the analysis is a recommendation that has a direct impact on the policies, practices, and investments the organization is likely to make with top management approval. All technical and business decisions are subject to challenges and may have to be altered in the future. Hence, the people originally involved in

Filled out by: Date:

1. Note the name of the evaluators and the date of evaluation.

Evaluated Aspect: Aspect of Activity/Product/Service (see form 3)
No. ____ Name ____________________

2. Go through your list of identified activities, products, and services and start with the first aspect of the first activity, product, or service. Note the aspect and the activity, product, and service on top of the form with the according number.

		3		2		1	
Area	**Factor**	**Value**	**Severe**	**Value**	**Moderate**	**Value**	**Low**
Environment	3		Severe impact		Moderate impact		Low impact
Legislative and other requirements	2		Essential including permits		Administrative or voluntarily		Not applicable
Management control	1		No control		Poor control		Controlled
Stakeholders internal/external	1		High importance		Medium importance		Low importance
					Overall result		0

3. Then go through the matrix of environmental criteria and mark down the respective weighting in the matrix for each aspect. The selected weighting is multiplied by the given factors. Document the total sum under Overall result.

Justification of weighting:

Environment:

Legislative and other requirements:

Management control:

Stakeholders internal/external:

4. Note how you came to your decision. Refer to supporting documents and information. In case you lack the necessary information note this down too. The data will be collected in the course of the Initial Environmental Review.

Determination of significance: Overall result 11 or higher Significant ___ Overall result under 11 Not significant ___

5. Finally, evaluate the significance of the reviewed aspect. If the overall result of the weighting is 11 or higher, the aspect is considered significant.

Figure 5–2 The methodology for identifying and setting priorities for significant environmental aspects.

the assessment are important because they may be future consultants or play a role in revisiting the recommendations at some future date. The age of the assessment is important because recommendations made today may not be appropriate or require updating several years down the road.

Step 2 involves reviewing a list of identified activities, products, and services. The list is developed after identifying the environmental aspects from the IER itself. The information documented in IER Auditor provides a baseline of the practices and regulatory status of the facility. From that baseline, an audit team must sit down and identify specific aspects and activities. The Refinery Material Output Matrices in Chapter 2 can also assist in narrowing down the aspects for specific unit processes within the manufacturing train. As an example, let us examine thermal cracking/visbreaking. Figure 5–3 summarizes the pollution output streams. This part of the refining process has two major pollution output streams: air emissions and process wastewater. In both streams, the sources of the pollution and causes can be identified. We can do the same exercise for all unit processes, operations, and other major operational aspects of the refinery. Thus, a list of evaluated aspects can be developed.

Step 3 is to evaluate whether the aspects are significant enough to warrant corrective action. Several criteria are used for such an

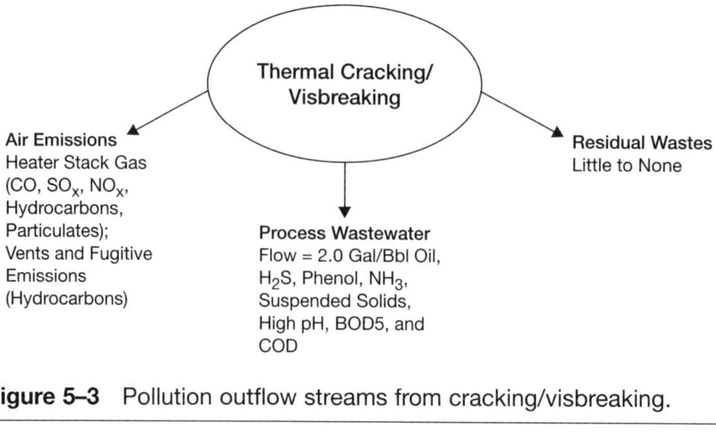

Figure 5–3 Pollution outflow streams from cracking/visbreaking.

assessment, but four general areas are often examined. Facilities focus on the direct impacts to the environment, whether or not the emissions or the control of the emissions meets regulatory (legislative) requirements and other standards, whether or not the output streams are under adequate management control; and finally, the output streams are assessed from the standpoint of their impacts to both internal and external stakeholders.

The reasons why a responsible company is concerned about negative aspects or interactions with the environment beyond meeting its regulatory obligations are obvious. Public health and safety and the preservation of natural resources and wildlife are reasons enough. Since these should be any facility's greatest priority, assessments of this category of aspects should be weighted more heavily than the other three. The aspects may be perceived as severe, moderate, or having low significance.

Meeting regulatory obligations is a necessity and a cornerstone of the EMS. Assessment of this category of aspects also should be weighted more heavily than the remaining two. A severe aspect would be one where it is essential to have permits or to meet other regulatory obligations. If those permits do not exist or have expired and the facility is discharging air emissions or liquid and solid wastes, then criminal negligence or willful disregard of the law are among the charges facility managers will face, for starters. A moderate impact would reflect that the aspect requires administrative or even a voluntary degree of control to act responsibly and reduce impacts. This can be a subject open for interpretation and perceived risks to a facility. For example, a facility might have a permit to discharge to the publicly owned treatment work (POTW) sewer system with no or minimal monitoring of the effluent and only a few parameters. Even though no regulations drive it to go beyond the minimum monitoring requirements, the facility may choose to do so, justifying the added costs and management from perceived future threats. Sewer lines break and some chemicals will leach and seep through sewer lines, thereby threatening the groundwater aquifer quality. Such contamination could migrate off-site and expose the facility to cleanup and civil actions from third-party damages. A good lesson can be learned from the wood treating industry, as a further example. In the early 1980s, it was an industry voluntary recommendation to install concrete drip pads around wood treating cylinders to minimize creosote, pentachlorophenol, and other wood treating chemicals from contaminating site operations during treatment operations. Not until the early 1990s did this become a regulatory requirement. Many facilities simply waited until they had a legal requirement to make the investment. Those companies that procrastinated until the compliance schedule was mandated faced both significant site cleanup challenges and, in a number of cases, third-party lawsuits from toxicity torts and property damages due to contaminated surface runoff problems. Related to the refinery industry itself, many gasoline retainers made the investments into meeting the technical upgrades for USTs years before they were legally required to meet a compliance schedule under RCRA Title C. That is not only acting responsibly, but it is proactive management of an environmental aspect. Finally, a low impact for this aspect category simply means that it is not applicable or has a very small potential impact based on the information available to the assessors.

Management control is an aspect category that surfaced as important in more recent times, particularly as EMSs became more fashionable throughout the latter part of the 1990s. The nature of an EMS forces better and more-attentive management control. But, in the beginning, significant deficiencies are found in the management of many environmental aspects. If this were not the case, then there would be no need for the EMS. A severe aspect reflects essentially no management control over the aspect. A moderate-level aspect reflects some level of control, but generally, it is poor. Finally, a low-level aspect is one where the aspect is controlled.

The last aspect-rating category is stakeholders. Internal stakeholders are employees, other operating divisions, marketing, R&D, and anyone or any operation potentially affected by the aspect being assessed. External stakeholders include investors, the public in terms of perceptions of corporate attitudes and practices toward the environment, banks, insurers, and many others. The exercise reduces to understanding how an aspect affects others and whether there are downsides to those relationships if the aspect is not controlled. A severe rating would reflect that the aspect is of high importance. A moderate rating reflects medium importance, and finally a low rating simply means the aspect is of low importance.

Since the auditors are faced with a number of aspects to evaluate and their task is to develop a recommendation as to which require immediate attention (i.e., a priority list of recommended actions), criteria are needed to efficiently rank the aspects relative to each other. To accomplish this, a numerical scoring system is used. This allows for a quantifiable setting of priorities as well as a measurable level of significance for each aspect evaluated. For a severe rating, values between 7 and 10 can be assigned based upon the audit team's consensus opinion. For a moderate rating of the aspect, values between 4 and 6 can be assigned, again based upon the group's consensus. And for a low rating, values between 1 and 3 can be assigned. Referring back to Figure 5–2, the ratings are entered into the evaluation table or matrix for the aspect under evaluation. Each number is then multiplied by the weighting factor (again a weighting fact of 3 is assigned to the Environment category, 2 is assigned to the Legislative category, and 1 is assigned to the Management Control and Stakeholder categories). By summing the multiplied values, a total score or value is obtained. The higher the score, the more significant the aspect. When the scores of several evaluated aspects are compared, those having the highest scores reflect operations with the most significant environmental aspects and, hence, should be addressed first. This forms the starting basis by which the team can begin devising formal strategy and action plans to recommend to management.

Returning to the methodology's sequence of steps outlined in Figure 5–2, Step 4 is to document the reasons for assigning the numerical score for each of the evaluation aspect categories. The assessment needs to be transparent and should stand up to peer review. Again, the analysis should result in a recommended strategy and action plans that senior management will consider. The decisions management makes carry with them financial and other resource allocations; hence, the logic behind or justification for assigning a level of significance for each evaluated aspect is important.

Step 5 has already been explained, but a few more words of explanation does not hurt. This step is the determination of the level of significance based on a numerical scoring system. The higher the numerical score, the more significant the aspect. When we repeat the exercise for several evaluated aspects and compare their values, the highest-priority ones become apparent. We also need to be sensitive that there exists a minimum threshold of significance, below which corrective actions are not warranted or simply not worth the effort. Since a major driver for investing in an EMS is the financial savings derived from improved environmental performance, an old adage that my father used to say comes to mind. If you walk through an operating plant and see a 50 cent piece lying on the floor, your natural inclination is to bend over, pick it up, and put it in your pocket. But most people will think twice about bending over to pocket the coin if the effort carries the risk of breaking your back. For this reason, a threshold "significance" value is assigned. Although somewhat arbitrary, based on my personal experience with hundreds of P2 audits, I have assigned a value of 11, below which the aspect is not deemed significant. Any score above 11 is significant but the aspect needs to be examined relative to other evaluated aspects.

This methodology has been programmed into a database management system that operates on Microsoft Access. The tool allows the reader to evaluate a large number of aspects, calculate a level of significance for each aspect, document the basis for each calculation, and sort the evaluated aspects according to their level of significance. The Tutorial contained as a subdirectory in EA Evaluator describes the operational features of the software.

5.4 Closing Remarks

Environmental management has become a big business over the past 10 years. An endless stream of consultants, consulting firms, pollution prevention specialists, and certified auditors compete for business and expound the need for acting responsibly through the application of ISO 14001 in particular. Many who consult and profess to be experts in the field are not technical but rather have straight management backgrounds. This is not necessarily bad, but it is not all that good. While it is true that an EMS is a management tool, its success within any organization hinges on the technologies implemented. One has to understand the technologies applied in manufacturing, whether a refinery, a fertilizer plant, or a machine-building operation. The limitations of the technologies and, in particular, the hardware contribute in a large way to the waste, inefficiencies, and pollution associated with any facility. These become compounded by operator error and inadequate training of workers. Because engineers are trained by education to solve problems, most can be placed into almost any manufacturing setting and engineer around the problems that impede efficient production, which in turn lead to cost efficiency. In short, you have to know and need to rely on technical solutions for any EMS to be successful.

It might surprise some readers to learn that the very same concepts we apply today, such as pollution prevention, good housekeeping practices, cleaner technologies, resource recovery, substituting less-toxic chemicals for more-hazardous ones, and waste minimization, were not introduced with the promulgation of the U.S. EPA in 1970. These practices predate the enforcement era by at least three decades. As an example, the following is a series of passages quoted from a 1939 book:[1]

- Handling of Chemicals—Plant and process layout should be designed to cut down handling of chemicals to the minimum possible. Corrosive liquids and solvents should be stored in closed and vented vessels and conveyed in closed pipe lines . . . bucketing should be avoided.
- Plant Design—Systematic inspection of processes and plant by full-time inspectors often reveals weakness in safety precautions. . . . A meeting of a committee on design is always called if major

[1]H. Tongue and G. Morgan, *A Practical Manual of Chemical Engineering* (London: Chapman and Hall, 1939), pp. 541–43.

alterations are desired, with the respect of the safety of the worker at the first thought.

- Protective Clothing and Appliances—A complete change of clothing, including underwear, is provided for certain processes, overalls for others, and rubber boots for most process workers. Goggles are scheduled for many applications. . . . Gloves are provided . . . rubber gloves alone are not permitted since they induce sweating and often contain chemicals of doubtful action to the skin. . . . Gas Masks and Respirators are always available and all staff are fully conversant with their use.
- Cleanliness—Housekeeping—baths—lavatories and the like. Special care must be taken in chemical works in regard to sanitary arrangements.
- Propaganda and Instructions—Every opportunity is taken of impressing workers with the necessity for their cooperation in regard to safety measures, and all staff fully and regularly instructed in accident prevention methods.
- Records should be kept of all accidents, and the occurrence of fires, so that remedial measures may be arranged on the result of periodical analysis thereof.

These guidelines are as fresh and applicable today as they were more than 60 years ago. They were added to and refined over the years and as knowledge and understanding of pollution problems created by industry activities began to be understood throughout the 1940s, 1950s, and the era of enforcement; more-detailed and more-rigorous standards were devised by the academic, engineering, and scientific communities and integrated with safety and environmental standards. To further illustrate this, the same 1939 book publication contains an appendix of equipment suppliers. There are vendors for equipment for air pollution controls, industrial water treatment systems, and evaporators for solvent recovery units.

Another historical example is a 1944 state-of-the-art review by Burke[2] for the modern chemical plant of the day. The review covers housekeeping improvements that affect on worker productivity and

[2]W. J. Burke, "Combating Health Hazards in the Chemical Industry, Part 2," *Chemical Industries* 55, October 1944: 565.

safety and preventive maintenance. These are low- to no-cost P2 investments. The review extends to discussions of process optimization improvements, optimizing existing processes that reduce resource consumption, and raw material substitution. Some environmental problems can be avoided by replacing hazardous materials with more environmentally benign materials. These options were recognized and identified as technical approaches to reduce waste and inefficiencies more than 60 years ago. The focus of this public-domain review was in adopting new technologies to reduce resource consumption and minimize waste generation through improved operating efficiencies. These same concepts are promoted under a modern-day EMS.

Starting in the 1920s, integrated oil companies paid ever-closer attention to waste and inefficiency because the pollution that leaked from wells, pipes, and tankers into the environment and leached from the unlined storage pits that dotted oil fields was product not sold to market. Gross pollution was bad for business. Indeed, in the early 20th century, perhaps one-fifth of the oil extracted from the ground never made it to market. This gave birth to the American Petroleum Institute (API), which focused on the creation of design and operating standards originally for refinery operations. But, since refinery operations use many of the same unit operations as chemical manufacturers, the standards and practices became universal to the chemical industry. The chemical industry, in turn, provided their customers with the same technical literature and standards. Since the 1920s, there have been over 900 API standards, which deal with a wide range of technical issues that have an impact on best practices and technologies. Some of these include spill prevention programs, application of tank dyking, proper tank venting, and the use of blowback systems and air emissions controls, water treatment operations, waste management, and stabilization techniques. These standards have continuously evolved and improved over the years. The API has been an organization to which all industry sectors have looked for best practices and recommended plant safety and environmental management practices for the entire history of the institute.

The point behind this stroll down memory lane is that industry on the whole has always known that it is not only in the public's best interest, but its own best interest to act in a responsible manner. The role of an EMS within any company is to pull the engineering talents under one orderly, structured system that tackles the wastes we refer to as *pollution*. Engineers per se do not make the best candidates to lead the organization's EMS efforts unless they had specific training in the environmental field and come from a strong management background. The fact that today the United States has more than 1200 Superfund sites, many of which stem from bygone days where engineers ruled the roost in plant management affairs is evidence enough that they are not the best candidates to take the commanding role. But, again, neither are individuals with straight management backgrounds and only an understanding of the compliance issues.

The choice of external consultants, auditors, and assigning responsibilities for implementing and enforcing the EMS from the talents within the organization requires serious considerations. If senior management is truly committed to the EMS, then it must seek out and pay a premium for the most-qualified personnel. Effort needs to be placed in developing a well-written job description, defining the qualifications that meet the needs of the organization. The credentials of the consultants and employees placed in roles of EMS responsibilities need to stand up to a high level of scrutiny.

Index